T0215949

Coexistence of IMT-Advanced Systems for
Spectrum Sharing with FSS Receivers in C-Band
and Extended C-Band

Lway Faisal Abdulrazak

# Coexistence of IMT-Advanced Systems for Spectrum Sharing with FSS Receivers in C-Band and Extended C-Band

 Springer

Lway Faisal Abdulrazak
Sulaimanyia Campus
Cihan University
Sulaimanyia, Iraq

ISBN 978-3-319-88965-8          ISBN 978-3-319-70588-0   (eBook)
https://doi.org/10.1007/978-3-319-70588-0

Printed on acid-free paper

This Springer imprint is published by Springer Nature
The registered company is Springer International Publishing AG
The registered company address is: Gewerbestrasse 11, 6330 Cham, Switzerland

*Dedicated to my parents, brothers, wife and beloved daughter and son*

# Abstract

Spectrum sharing between wireless systems becomes a critical issue due to emerging new technologies. In 2007, the International Telecommunication Union (ITU) recommends spectrum sharing feasibility to be established for IMT-Advanced systems and fixed satellite services (FSS) for a frequency range 3400–4200 MHz. This book presents a research conducted on the interference mitigation between IMT-Advanced and FSS. It covers a deterministic analysis for interference-to-noise ratio (I/N), adjacent channel interference ratio (ACIR), field strength and path loss propagation, in order to determine the separation distances in the co-channel interference (CCI) and adjacent channel interference (ACI) scenarios. An analytical model has been developed for the shielding mitigation technique based on the deterministic analysis of the propagation model. The shielding technique has been developed based on test bed measurements for evaluating the attenuation of the proposed materials. MATLAB™ and Transfinite Visualyse Pro™ have been used as simulation tools for the verification of the obtained results, whereas the IMT-Advanced parameters have been represented by Worldwide Interoperability for Microwave Access (WiMAX) 802.16e. The impact of different FSS channel bandwidths, guard band separations, shielding effects, antenna heights and different deployment areas on coexistence feasibility is considered. The results obtained in terms of minimum required separation distance in three scenarios, co-channel, zero-guard band and adjacent channel, are analysed. In order to reduce the emitted interference from the IMT-Advanced base station, the improved multiple signal classification (I-MUSIC) algorithm is derived. It has a high resolution and accurate direction of arrival detection and prevents the high heavy complexity of current methods. Finally, a mechanism system for adaptive array antenna is designed to keep fixed nulls in the direction of FSS and steer the beam towards the users by employing the I-MUSIC and the least mean square (LMS) algorithms. The efficiency of proposed mitigation technique is justified using simulation results, and coexistence is characterized by minimum separation distances. It has been found that 20 dB attenuation is needed in order to reduce the original separation distance by 90%. Moreover, the optimum separation distance reduction achievable by using 40 dB shielding and fixed null mitigation technique could reach up to 0.335% of the original separation. Accordingly, the shielding and proposed adaptive antenna mitigation techniques have demonstrated a high degree of capability and competence of reducing the harmful power interference from the IMT-Advanced base station to the FSS receiver.

# Contents

# List of Abbreviations

| | |
|---|---|
| 2D | Two dimensional |
| 4G | Fourth generation |
| ACIR | Adjacent channel interference ratio |
| ACS | Adjacent channel selectivity |
| AoA | Angle of arrival |
| APT | Asia Pacific Telecommunication |
| AWGN | Additive white Gaussian noise |
| BiSVD | Bi-iteration singular-value decomposition |
| BS | Base station |
| BW | Bandwidth |
| BWA | Broadband wireless access |
| CCI | Co-channel interference |
| CEPT | The European Conference of Postal and Telecommunications Administrations |
| C/I | Carrier-to-interference ratio |
| C/N | Carrier-to-noise ratio |
| C/N+I | Carrier-to-noise plus interference ratio |
| CMA | Constant modulus algorithm |
| DAMA | Demand assigned multiple access |
| DOA | Direction of arrival |
| EIRP | Effective isotropic radiated power |
| ES | Earth station |
| ETSI | The European Telecommunications Standards Institute |
| EVD | Eigenvalue decomposition |
| EMI | Electromagnetic incompatibility |
| ESPRIT | Estimation of signal parameters via rotational invariance technique |
| FCC | Federal Communications Commission |
| FS | Fixed services |
| FSS | Fixed satellite services |
| FWA | Fixed wireless access |

| | |
|---|---|
| GB | Guard band |
| GOS | Geostationary orbit station |
| I-MUSIC | Improved MUSIC |
| I/N | Interference-to-noise ratio |
| IA | Interference area |
| IDU | Indoor unit |
| IMT | International Mobile Telecommunications |
| IMT-2000 | International Mobile Telecommunications-2000 |
| IMT-Advanced | International Mobile Telecommunications-Advanced |
| ISOP | Interference scenario occurrence probability |
| ITU | International Telecommunication Union |
| ITU-R | International Telecommunication Union Radiocommunications Sector |
| LCMV | Linearly constrained minimum variance filter |
| LMS | Least mean square |
| LNB | Low-noise blocking |
| LOS | Line of sight |
| LP | Linear prediction |
| LS-ESPRIT | Least squares-estimation of signal parameters via rotational invariance technique |
| MC | Monte Carlo |
| MCL | Minimum coupling loss |
| MCMC | Malaysian Communications and Multimedia Commission |
| MIMO | Multiple input, multiple output |
| MMSE | Minimizing mean square error |
| MSN | Maximizing signal-to-noise ratio |
| MUSIC | Multiple signal classification |
| NLOS | Non-line of sight |
| NMS | Network management system |
| OFCOM | The Federal Office of Communication |
| OFDM | Orthogonal Frequency Division Multiplexing |
| OFTA | Office of the Telecommunications Authority |
| OOB | Out-of-band emission |
| ODU | Outdoor unit |
| QPSK | Quadrature phase-shift keying |
| RF | Radio frequency |
| Rx | Receiver |
| SEM | Spectral emission mask |
| SINR | Signal-to-interference plus noise ratio |
| SNOIs | Signal not of interest |
| SOI | Signal of interest |
| SS | Subscriber station |
| TLS-ESPRIT | Total least squares-estimation of signal parameters via rotational invariance technique |
| TVRO | Television receive-only |

| TX | Transmitter |
|---|---|
| ULA | Uniform linear array |
| VSAT | Very small aperture terminal |
| WiMAX | Worldwide interoperability for microwave access |
| WINNER | European Wireless World Initiative New Radio project |
| WLL | Wireless local loop |
| WP | Working party |
| WRC | World Radiocommunication Conference |

# List of Appendices

# List of Figures

# List of Tables

# Chapter 1
# Introduction

## 1.1 Research Background

The frequency spectrum is a precious natural resource, subject to rapid consumption and necessary for global, regional and domestic telecommunication infrastructures [1, 2]. In the World Radio Conference WRC-07, the International Telecommunication Union for Radiocommunication (ITU-R) has become the key regulator of the global spectrum allocation for the next generation of mobile systems [3]. In this sense, this book addresses the (3400–4200) MHz band of the spectrum, which has been proposed by the ITU-R as the widest band that will be available, up to 100 MHz/channel, for the future International Mobile Telecommunication Advanced (IMT-Advanced) operational frequency. For fixed satellite services (FSS), C-band is used in many countries since 1980, represented by thousands of strategic investments ranging from telemedicine and distant learning to disaster recovery [4]. Accordingly, any immediate transition in the use of this band to IMT-Advance services is considered unrealistic [5].

The super extended C-band 3400–4200 MHz is attractive for FSS because of its low absorption, highly reliable space-to-earth communication and wide service coverage. In addition, this frequency band is widely used by satellite operators in the countries with severe rain fade conditions due to almost zero rain-induced signal attenuation. Moreover C-band is also favourable to IMT-Advanced, because it allows multiple antenna technique implementations and the use of smaller antenna for terminals and base stations, as well as enabling high space efficiency [4].

In response, the frequency administrators explored the expected interference between the FSS and IMT-Advanced. Studies, analyses and measurements have been reported since 2005 to improve the efficiency of receiving signals via FSS. Furthermore, various studies have been conducted on similar cases of interference between the terrestrial communications like the fixed wireless access (FWA)

© Springer International Publishing AG 2018
L.F. Abdulrazak, *Coexistence of IMT-Advanced Systems for Spectrum Sharing with FSS Receivers in C-Band and Extended C-Band*,
https://doi.org/10.1007/978-3-319-70588-0_1

and FSS, due to the relative higher power of the FWA signal comparing to the satellite signal [6]; details are carried out in Chap. 2.

The co-channel interference (CCI) and adjacent channel interference (ACI) are issues that result of co-locating more than one service in one band. However, CCI is the worst of the issues in the coexistence of both IMT-Advanced and FSS using the same frequency. Overcoming this problem is the subject of utilizing geographical domain separation. ACI results from other signals that are adjacent in the frequency to the desired signal. Using the adjacent band for different services side by side (band segmentation) with or without guard band (GB) could be a possible solution for ACI [7].

The sharing results by using a minimum coupling loss (MCL) and Monte Carlo (MC) simulation link gave a required separation distance larger than 40 km to avoid mutually harmful interference between two systems in co-channel and adjacent channel interference scenario [8, 9]. MCL is the minimum possible propagation loss between station and non-coordinated operators, neglecting the effect of lognormal fading (shadowing fade) and MC is used to quantify the minimum frequency separation of mobile communication system in adjacent frequency band [9]. On the IMT-Advanced side, the orthogonal frequency division multiplexing (OFDM) is currently considered the most promising access schemes to support IMT-Advanced systems [10]. It is based on multi-carrier modulation technique that offers excellent performance in combating multipath fading as well as superb efficiency in terms of using the available bandwidth [11].

From the radio propagation standpoint, ITU-R has specified propagation environments for evaluating transmission performance for same emerging wireless technologies [12]. These include both terrestrial and satellite propagation situations. It is concluded that IMT-Advanced operating environments are dense urban, urban, suburban and rural, with several common characteristics, such as interference to noise ratio (I/N).

The C-band is heavily congested due to the following: first, such countries as Japan, Korea and Australia with highly developed telecommunications infrastructure need a new spectrum for high-capacity mobile broadband wireless access (BWA) networks; second, countries (such as China, India, Australia) with large underserved rural areas need new spectrum for fixed BWA networks [13].

From the literature, it is concluded that an easy-to-follow approach is needed in order to determine the frequencies' coordinations with different bandwidths, the focus of this book.

## 1.2  Problem Statement

The coexistence between IMT-Advanced systems and the FSS receiver in C-band lower frequency is the main focus of this book. The 3400–4200 MHz frequency band, previously allocated for existing FSS receivers, has been proposed to be investigated for IMT systems use. Among the many built-in FSS receiver capabilities, exceptional weak signal received from geostationary orbit station (GOS) is perhaps the unique feature inherent to that kind of receivers. Therefore, a potential interference from other systems into FSS receivers is inevitable. However, it is difficult to identify a fixed interference level for all FSS receivers due to different assigned bandwidths. These assigned bandwidths start from 230 kHz to 100 MHz for some geostationary satellite orbit (GSO). On the mobile-terrestrial side, IMT-Advanced signals should be robust enough to support the so-called interoperability supported by various services. Therefore, interference between IMT-Advanced and FSS services is likely to be in one side, which is from IMT-Advanced (interferer) to FSS (victim). This is the unsolved problem which has motivated this research work.

The current techniques of mitigating the impact of IMT-Advanced based on the separation distances are not logical due to distance infeasibilities. Thus, it is important to investigate the FSS interference thresholds for different bandwidths in order to validate the effectiveness of the proposed mitigation techniques on the separation distance reduction. In addition, field measurements are required in order to get accurate interference predictions, which are currently unavailable for IMT-Advanced systems. Therefore, it is useful to develop an analytical model for the FSS shielding protection by measuring the exact value of shielding attenuation for different selected materials.

The coexistence model between IMT-Advanced represented by Worldwide Interoperability for Microwave Access (WiMAX) 802.16e and FSS receiver is proposed to study the impact of different FSS channel bandwidths, guard band separations, shielding effects, WiMAX transmitting antenna heights and different deployment areas, which may likely have an influence on coexistence feasibility.

Technically, it is impossible to achieve a feasible separation distance in most cases without nulling the WiMAX power in the direction of FSS and increase the signal to noise ratio (SNR) towards the signal of interest. Accordingly, developing an algorithm, which can operate with a low SNR to extract fixed nulls towards the FSS and form the beams towards IMT-Advanced users, will be a possible mitigation technique to reduce the separation distance.

## 1.3   Research Objective

The main objective of this research is to develop an analytical coexistence model which can predict the minimum separation distance between IMT-Advanced and FSS. This model will mainly depend on the usage of mitigation techniques that involve parameters such as the guard band, shielding and IMT-Advanced fixed null extraction. The research objectives are summarized as follows:

- To investigate the required separation distance and guard band using variable FSS heights, bandwidths and different deployment scenarios

- To measure the amount of the shielding attenuation through different proposed materials and suggest the most feasible shielding technique to protect the FSS

- To improve the MUSIC algorithm in order to increase the direction of arrival (DOA) resolution for the signal of interest

- To design a null extraction mechanism using uniform linear array (ULA) in order to form the beam using least mean square (LMS) algorithm, while nulls are fixed towards FSS and DOA will be determined using developed MUSIC algorithm

## 1.4   Scope of Work

As the band 3400–4200 MHz is allocated worldwide on a primary basis to the fixed satellite service (FSS) [14], this research will investigate the possible mitigation techniques that coexist between the FSS receiver using MEASAT-3 GSO (channel bandwidth 36 MHz maximum) and IMT-Advanced systems represented by WiMAX 802.16e (channel bandwidth 20 MHz). The results of this study will not consider the unevenness of earth surface. The present study is designed to determine the benefits of using higher frequency for IMT-Advanced system. This study will be carried out in the same and adjacent frequency band. The scope of the research has been listed as follows:

- Literature reviews have been carried out on frequency bands, coexistence and sharing criteria, spectrum allocation, interference types and concept, signal propagation, adjacent channel leakage ratio (ACLR) model, smart antenna systems, shielding and null extraction technique and previous related studies.

- Clarifying the parameters and specifications of IMT-Advanced (proposed system WiMAX IEEE802.11e) and FSS receiver. Then, determine the interference level for each FSS bandwidth and consider the most used bandwidths for the analytical model.

- Identifying the radio wave propagation formulas in order to find the minimum separation distances in the CCI and ACI scenarios.

- Development of an analytical model that is capable of enrolling the ACIR model, shielding and null extraction techniques into the minimum separation distance analysis.

- Measure the received signal by FSS via MEASAT-3 and evaluate the effect of different shielding materials to conclude the best shielding material and scenario.

- Apply a terrestrial signal generated by signal generator (that has the same frequency carrier of victim FSS) to assess the interference with and without shielding.

- Determine the intersystem interference scenarios in different deployment areas according to the requirements and criterion by simulation.

- Present the problem of high resolution and accurate direction of arrival detection using Improved MUSIC (I-MUSIC) algorithm. Improving the existing MUSIC spectrum model by scanning in only one dimension. This can prevent the high heavy complexity of current methods. Then, the results will be verified by simulation. Here, I-MUSIC is driven which distinctively reduces the complexity of DOA estimation. In addition, comparisons between the root mean squared error (RMSE) and computational complexity costs against those of conventional algorithms are performed. Numerical results for different array antenna manifolds and a variety of data lengths are also presented in the simulations.

- Propose a mechanism for beam cancellation to coexist the IMT-Advanced base stations (BS) and FSS in the 3400–4200 MHz frequency range. This mechanism should be able to steer the radiation power towards the user while still fixing the fixed nulls in the direction of FSS receiver.

Finally, the most feasible coexistence scenarios will be determined, and the feasibility of the proposed mitigation techniques on the coordination process will be evaluated.

## 1.5   Book Significance and Contribution

In this book, the author describes the techniques of IMT-Advanced and fixed satellite services sharing the same spectrum and presents a number of original contributions in this field of study. These contributions are summarized as follows:

- A significant study based on measurements and experimental setup has been performed to improve and mitigate the impact of terrestrial services on the FSS receiver.

- Different scenarios in the co-channel, zero-guard band and adjacent channel, with 0 dB and 20 dB shielding attenuation for each, have been proposed to obtain the correlation between the minimum separation range of base stations and the frequency separation. A propagation model is derived for this purpose. The model can predict the systems coexistence possibility in several deployment areas. Besides, different FSS channel bandwidths had been considered to further substantiate the results in terms of various applications of satellite receivers.

- I-MUSIC algorithm has been proposed as a competent algorithm for high-resolution DOA detector.

- A competent mechanism to steer the radiation pattern towards IMT-Advanced users and keep the nulls towards FSS victim is presented in this book. It is found that this technique can significantly reduce down the separation distance to

35.5%. It has been simulated for the co-channel and adjacent channel interference scenarios, in different shielding attenuation and different deployment areas.

- The results provided in this research have shown that frequency sharing between FSS and IMT-Advanced systems in 3400–4200 MHz bands is feasible under certain conditions. Accordingly, shielding and fixed null extraction mitigation techniques have been used for a high competence and capability to reduce down the interference to 0.355% of the original.

So far, broad and thorough research activities have been conducted to study the anticipated deteriorating effects of interference between FSS and IMT-Advanced system.

## 1.6  Book Outlines

This book is organized in six chapters to cover the whole research work that has been conducted.

The second chapter provides a summary of literature review on radio propagation and the deterministic analysis for the CCI and ACI. It includes results of the most recent studies, assessments and interference method developments. The concept, effect and types of shielding are discussed in detail. The new technology of smart antenna parameters using MUSIC algorithm, as well as its background and concept, is highlighted.

The third chapter proposes a methodology of an interference model in the co-channel and adjacent channel interference scenarios, which essentially depends on the interferer adjacent channel leakage ratio (ACLR), receiver adjacent channel selectivity (ACS) and the clutter loss effect. A case study of one channel of MEASAT-3 C-band is considered for initial planning and frequency coordination on WiMAX of 20 MHz channel bandwidth. Frequency offsets and geographical separations for different deployment environments are considered. Based on interference to noise ratio (I/N) of −10 dB, the calculations are performed for 4 GHz frequency carrier. Matlab code has been developed so that the specified value of I/N is obtained by tuning up the minimum required separation distance to an appropriate level corresponding to a set of frequency offsets between carriers.

Chapter 4 describes the details of an experimental setup in Universiti Teknologi Malaysia, Skudai, Johor, and an analytical model for shielding technique. The experimental setup comprises of FWA base station installation and FSS receiver setup. Compatibility between IMT-Advanced system represented by WiMAX 802.16e and fixed satellite service system will be investigated for both co-channel system base stations and adjacent channel system base stations. Assessment of the interference from WiMAX to FSS will be performed in terms of guard band, antenna discrimination, shielding, different FSS bandwidth and diverse deployment areas to achieve minimum separation distance for each scenario.

An efficient method for the pattern synbook of the linear antenna arrays with the prescribed null is presented in Chap. 5. The DOA algorithm, which can give a high resolution, is proposed. Then, adaptive beamforming algorithm based on Improved MUSIC algorithm is combined with the LMS algorithm to handle adjustable code for null construction in the direction of the victim FSS earth station (ES). An assessment will be done for the interference modes where the interfering signal emitted from one IMT-Advanced BS impacts one FS station.

In the sixth chapter, the overall conclusions of research work conducted under this research are presented. Finally, the recommendations for future work related to IMT-Advanced physical layer are presented.

# References

1. D. Laster, J.H. Reed, Interference rejection in digital wireless communications. IEEE Commun. Mag **14**, 37–62 (1997)
2. Z.A. Shamasn, L. Faisal, T.A. Rahman, On coexistence and spectrum sharing between IMT-advanced and existing fixed systems. Int. J. Publ. WSEAS Trans. Commun 7(5), 505–515 (2008)
3. H.-G. Yoon, W.-G. Chung, H.-S. Jo, J. Lim, J.-G. Yook, H.-K. Park, Spectrum requirements for the future development of IMT-2000 and systems beyond IMT-2000. (First published on 2006 and updated on 2012). J. Commun. Netw. **8**(2), 169–174 (2006). https://doi.org/10.1109/JCN.2006.6182744
4. ITU-R Document 8F/1015-E, *Sharing Studies Between FSS and IMT-Advanced Systems in the 3400–4200 and 4500–4800 MHz Bands*, (2006). http://portal.acm.org/citation.cfm?id=1456088
5. ITU-R WP 8F/TEMP 432 rev.2, *Working Document Towards a PND Report on Sharing Studies Between IMT-ADVANCED and the Fixed Satellite Service in the 3400–4200 and 4500–4800 MHz Bands*. ITU-R Working Party 8F, August 2006. http://www.itu.int/publ/R-REP-M.2109/en
6. L. F. Abdulrazak, T. A. Rahman, Review ongoing research of several countries on the interference between FSS and BWA. *International Conference on Communication Systems and Applications (ICCSA'08)*, Hong Kong China, 7–9 Sept 2008. pp. 101–108
7. IST-4-027756 WINNER II D 5.10.1, *The WINNER Role in the ITU Process Towards IMT-Advanced and Newly Identified Spectrum*, Vol.0, November 2007
8. H.-S. Jo, H.-G. Yoon, J.W. Lim, J.-G. Yook. An advanced MCL method for assessing interference potential of OFDM-based systems beyond 3G with dynamic power allocation. *Proceedings of the 9th European Conference on Wireless Technology*. UK. Sept 2006. pp. 39–42
9. P. Seidenberg, M.P. Althoff E. Schulz, G. Herbster, M. Kottkamp, Statistical of minimum coupling loss in UMTS/IMT-2000 reference scenarios. *Vehicular Technology Conference VTC'99, IEEE VTS 50th*. Vol:2, Publication Year: 1999, pp. 963–967
10. ITU-R WP8F Contribution, *Proposed MIMO Channel Model Parameters for Evaluation of Air Interface Proposals for IMT-Advanced*. Document 8F/1149-E, Question ITU-R 229/8, ITU-R WP8F Meeting in Cameroon, January 2007. http://www.itu.int/md/meetingdoc.asp?lang=en&parent=R03-WP8F-C&PageLB=225
11. X. Li, L.J. Cimini Jr., Effects of clipping and filtering on the performance of OFDM. IEEE Commun. Lett. **2**, 131–133 (1998)

12. Recommendation ITU-R P.452-12, *Prediction Procedure for the Evaluation of Microwave Interference Between Stations on the Surface of the Earth at Frequencies Above About 0.7 GHz*, Geneva, Switzerland, May 2007
13. L.F. Abdulrazak, Z.A. Shamsan, T.A. Rahman, Potential penalty distance between FSS receiver and FWA for Malaysia. Int. J. Publ. WSEAS Trans. Commun.. ISSN: 1109-2742 **7**(6), 637–646 (2008)
14. L.F. Abdulrazak, A.A. Odah, Tractable technique to evaluate the terrestrial to satellite interference in the C-band range. Int. J. Theor. Appl. Inform. Technol **65**(3), 762–769 (2014)

# Chapter 2
# Literature Review

## 2.1 Introduction

Broadband wireless access (BWA) systems have been introduced to operate in all or portions of the 3400–4200 MHz band for fixed, nomadic or mobile user terminals. As BWA is being introduced, harmful interference and loss of FSS receivers have been reported in Malaysia [1]. The reported cases cover interference both for BWA in overlapping frequency bands and in non-overlapping bands. In the ITU table of frequency allocations, the FSS, in the space-to-earth direction and the Fixed Service (FS), is co-primary in the band 3400–3800 MHz. In some national tables of frequency allocations, the FSS is not primary in the band 3400–3700 MHz or over a portion of this 300 MHz range. Currently FSS is used over the whole 800 MHz range, but the utilization of the upper 500 MHz (3700–4200 MHz) is much more intense, followed by the utilization of the 3625–4200 MHz band [2–4]. Therefore in this chapter, spectrum sharing studies will target the principles of radio propagation and the types and criteria of intersystem interference. A literature review of previously conducted studies, assessments and interference methods will be discussed in this chapter. In addition, the background and concept of shielding technique, smart antenna elements and MUSIC algorithm will be highlighted and discussed in detail as a mitigation technique.

## 2.2 Radio Propagation

In wireless communications, radio propagation between base station and terminals is affected by such mechanisms as scattering, diffraction and reflection.

© Springer International Publishing AG 2018
L.F. Abdulrazak, *Coexistence of IMT-Advanced Systems for Spectrum Sharing with FSS Receivers in C-Band and Extended C-Band*,
https://doi.org/10.1007/978-3-319-70588-0_2

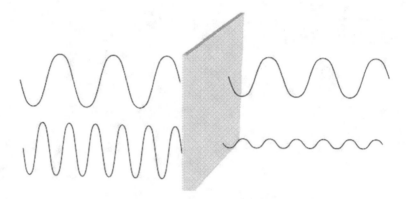

**Fig. 2.1** Higher frequencies have higher attenuation on penetrating obstacles

## 2.2.1  Radio Coverage

The radio coverage is determined by radio signal path loss, which increases with increasing frequency. The RF power of radio signals would be reduced when radio signals have travelled over a considerable distance. Therefore, in most cases, the systems with higher frequencies will not operate reliably over the distances required for the coverage areas with varied terrain characteristics [5]. For clear line-of-sight (LOS) propagation, the range between the transmitter and receiver is determined by the free space path loss equation, given by:

$$\text{Pathloss} = 20\log_{10}\left(\frac{4\pi d}{\lambda}\right)\text{dB} \tag{2.1}$$

where $d$ and $\lambda$ are the range and wavelength in meters, respectively.

In non-line-of-sight (NLOS) cases, the performance of higher frequencies is worse with reliable distances dropping even faster. Most paths are obstructed by objects and buildings. When penetrating obstacles, radio waves are decrease in amplitude. As the radio frequency increases, the rate of attenuation increases. Figure 2.1 illustrates the effect of higher frequencies having higher attenuation on penetrating obstacles [6].

A radio beam can diffract when it hits the edge of an object. The angle of diffraction is higher as the frequency decreases. When a radio signal is reflected, some of the RF power is absorbed by the obstacle, attenuating the strength of the reflected signal. Figure 2.2 shows that higher frequencies lose more signal strength on reflection [7].

Conversely, high frequency is required to provide sufficient bandwidth. However, spectrum allocation widths are normally proportional to the frequency of the band, and hence nominating the 3400–4200 MHz band for IMT-Advanced would allow the spectrum users to operate with more and wider channels.

The use of higher available capacity can also support much higher data rates than the lower spectrum. In addition, higher frequency can reduce the financial cost of

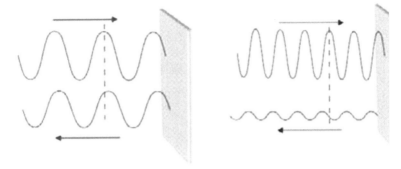

**Fig. 2.2** Frequency dependence of signal strength on reflection

licensing. It is important to notice that gain of antennas is a functional to the frequency being received [8].

In free space propagation, clear and unobstructed line-of-sight (LOS) path is available, and the first Fresnel zone is maintained between base station and terminal. Free space path loss can be obtained by using the logarithmic value of the ratio between the receiving and transmitting power as expressed in Eqs. (2.2), (2.3), (2.4) and (2.5). This simplified free space path loss model for unity antenna gain is based on Ref. [9].

Equations (2.3), (2.4) and (2.5) indicate that free space path loss is frequency dependent, and it increases with distance. The increase of distance and frequency produces similar effect on the path loss.

$$PL_{dB} = 10\log_{10}\frac{P_r}{P_t} \tag{2.2}$$

$$PL_{dB} = -147.56 + 20\log_{10} f_{Hz} + 20\log_{10} d_m \tag{2.3}$$

$$PL_{dB} = 32.44 + 20\log_{10} f_{MHz} + 20\log_{10} d_{km} \tag{2.4}$$

$$PL_{dB} = 92.44 + 20\log_{10} f_{GHz} + 20\log_{10} d_{km} \tag{2.5}$$

where $f$ is frequency, $d$ is distance; $P_r$ and $P_t$ are the receiving and transmitting power in watts, respectively.

## 2.2.2   Radio Propagation Model

A radio propagation model is an empirical mathematical formulation for the characterization of radio wave propagation as a function of frequency, distance and other characteristics. A single model is usually developed to predict the behaviour of propagation for every similar link under similar constraints. The essential aim of

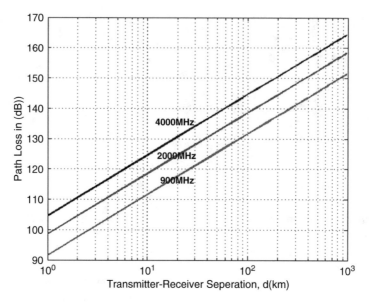

**Fig. 2.3** Free space path loss at 900, 2000 and 4000 MHz

signal propagation studies is to formalize how the signal can propagate from one point to another. Only in such situation can a typical model predict the path loss effect on an area covered by a single- or multi-transmitter (s) [10].

It is found that ITU-R P.452-9 [11] is the most suitable propagation model for this study, because it can cover from 0.7 MHz to 30 GHz frequency range. The prediction of the line of sight (LOS) is a result of the signal after being exposed to the path and clutter loss model, as clarified in Eq. (2.6). The propagation prediction can also be obtained by summing line of sight with subpath diffraction and clutter model, as in Eq. (2.7).

$$L_b = Pl_{\text{free}} + A_{\text{ht}} + A_{\text{hr}} \tag{2.6}$$

$$L_b = Pl_{\text{free}} + L_{\text{ds}} + A_{\text{ht}} + A_{\text{hr}} \tag{2.7}$$

where $L_b$ is the prediction basic transmission loss given by the line-of-sight model, while $A_{\text{ht}}$ and $A_{\text{hr}}$ represent the propagation losses encountered due to the different heights in one environment and $L_{\text{ds}}$ is a result of diffraction loss from prediction of subpath loss obtained from the diffraction model. An analytical predication algorithm for the horizontal transmission can be expressed as:

$$L_b = -5\log\left(10^{-0.2L_{\text{bs}}} + 10^{-0.2L_{\text{bd}}} + 10^{-0.2L_{\text{ba}}}\right) + A_{\text{ht}} + A_{\text{hr}} \tag{2.8}$$

where $L_{\text{bs}}$, $L_{\text{bd}}$ and $L_{\text{ba}}$ are individually predicted basic transmission loss obtained by troposcatter, diffraction and ducting layer reflection propagation models, respectively.

CEPT and ITU organizations have accepted a common formula for wireless transmission assessment at a microwave frequency level. This formula has incorporated the clutter attenuation as well as environmental effects and is expressed as follows:

$$L(d) = 92.44 + 20\log d + 20\log f + A_h \qquad (2.9)$$

where $d$ (km) is the distance between interferer and victim FSS receiver, $f$ is the carrier frequency in GHz and $A_h$ is loss due to protection from local clutter (i.e clutter loss) and is given by:

$$A_h = 10.25e^{-d_k}\left[1 - \tanh\left[6\left(\frac{h}{h_a} - 0.625\right)\right]\right] - 0.33 \qquad (2.10)$$

where $d_k$ (km) is the distance from nominal clutter point to the antenna, $h$ is the antenna height (m) above local ground level and $h_a$ (m) is the nominal clutter height above local ground level. In Ref. [11], clutter losses are evaluated for different categories, such as trees, rural, suburban, urban and dense urban. Increasing antenna height up to the clutter height will result in a decrease in clutter loss, as shown in Table 2.1 and Fig. 2.4.

Table 2.1 reveals that the value of nominal distance is highest for rural and suburban areas, whereas for urban and dense urban areas, the separation distance decreases in response to the clutter loss increment. Nominal distance is associated with the loss issues due to clutter height, which depends on the deployment areas. The detailed analysis is explained in Chap. 3.

**Table 2.1** Nominal clutter heights and distances [11]

| Clutter category | Clutter height $h_a$ (m) | Nominal distance $d_k$ (km) |
| --- | --- | --- |
| Rural | 4 | 0.1 |
| Suburban | 9 | 0.025 |
| Urban | 20 | 0.02 |
| Dense urban | 25 | 0.02 |

## 2.3  Coexistence, Spectrum Sharing and ITU Allocation in 3400–4200 MHz Band

Sharing a particular frequency among users using different devices has become the most critical problem in wireless communication systems. In order to guarantee a smooth access to the wireless channels, it is proposed to allocate spectrum using sharper frequency masks [12]. Wireless services have been developed in multi-deployment forms for many applications in recent years. This development has

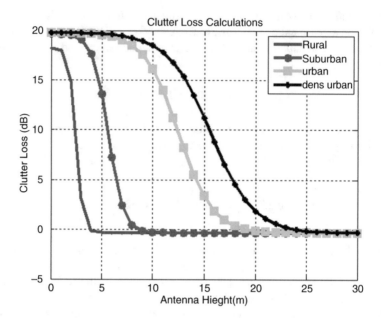

**Fig. 2.4** Clutter losses for rural, suburban, urban and dense urban areas

made the frequency bands become limited and highly scarce. Therefore, it is very important for emerging systems to be able to share and coexist with other systems on either co-channel or adjacent channel [13].

Spectrum coexistence means that different devices can have access to the spectrum in the same or adjacent frequency band(s) without causing any interference. This is possible by setting several regulatory rules regarding separation distance, frequency separation and power transmission. The coexistence studies emphasize on the performance of systems when they operate in the same or adjacent frequency bands [14, 15].

Spectrum sharing is defined as the use of a same frequency band by different systems or services, either with system coordination or not [13]. Sharing of the same frequency band by different services or technologies is only possible through well-defined limitations and technical requirements which facilitate sharing capabilities.

Table 2.2 shows the services in the 3400–4200 MHz range in the three regions [16]. The primary service is denoted by capital letters, while possible secondary services are denoted by lower-case. The secondary service is allowed to operate if no interference is caused to the primary service.

**Table 2.2**  ITU spectrum allocation in 3400–4200 MHz (the targeted band)

| Band (GHz) | Region 1 | Region 2 | Region 3 |
|---|---|---|---|
| 3.4–3.5 | Fixed, fixed satellite (space-to-earth), mobile, radiolocation | Fixed, fixed satellite (space-to-earth), amateur, mobile, radiolocation | |
| 3.5–3.6 | | Fixed, fixed satellite (space-to-earth), mobile (apart from aeronautical), radiolocation | |
| 3.6–3.7 | Fixed, fixed satellite (space-to-earth), mobile | | |
| 3.7–4.2 | | Fixed, fixed satellite (space-to-earth), mobile (apart from aeronautical) | |

## 2.4   Fixed Satellite Services Deployment Topology and System Design

Satellite services in C-band are essential for inhabitants of rural remote areas as well as tropical weather areas. As a solution, a very small aperture terminal (VSAT) associated with a local loop is employed. Typically, in these environments 200–500 telephone lines are required for each local loop [17].

VSAT is normally a small dish used for satellite communication services. It can offer a reliable and supportable space-to-earth and earth-to-space signal transmission solutions. It is important for both national and international services, on land and at sea. VSATs are most effective where the existing telecommunication infrastructures are unreliable. However, even in developed countries, VSAT can provide effective data distribution at highly competitive costs as the terrestrial networks [18].

Usually the FSS connection scenario in C-band is a combination of multiple access infrastructures and the wireless local loop (WLL). System design considerations include the type of traffic which can be carried, traffic rate, VSAT network architecture and topology. Others are modulation scheme used, multiple access schemes used, footprint, link budget and ground segment antenna design. The VSAT networks provide:

1. Reliable data transmission with different bandwidths and adaptive power. These data can be voice and video transmitted to different users within the coverage area with ease of network management tools.
2. Broadcasting services (point-to-multipoint), data collection (multipoint-to-point), broadband and point-to-point communications are a part of VSAT applications.
3. Interactive real-time applications, telephony, Internet, multimedia delivery and direct-to-home.
4. VSAT is applicable in all land areas, remote locations, water areas and large volumes of air space.

In the satellite industry, VSAT term is used to describe the earth station unit used for transmitting to or receiving electromagnetic signals from satellite or aircraft [19].

In terms of topology, the FSS can have three forms of connections. The first one is the star, which the hub controls the data transmission between users. A large number of users in this case can have access to the network. The hub antenna diameter ranges between 6 m and 11 m. A centralized data application form is what this topology is suitable for, because all users can communicate with a single hub. The second type is mesh connection. In this case, a group of VSAT users can communicate directly among the network terminals with less time delay, which makes it very attractive for telephony applications. The third topology is hybrid network, which is a mixed network of mesh and star topology [17]. The received signal by FSS is the output of transmitted signal from the GSO, transponder antenna gain and receiving FSS gain. It is important to consider the free space and equipment losses. Earth station hub has higher gain and diameter (4–11 m), and it communicates with all terminals and requires higher bandwidth.

## 2.5   Vision for IMT-Advanced System Concept

The growth of International Mobile Telecommunications-2000 (IMT-2000) is anticipated to reach a limit of around 30 Mbps [20]. Then, IMT-Advanced is expected to be the next generation of the mobile communication system. As clarified by the ITU, the system has specific greater capabilities than that of IMT-2000 [20].

Currently, ITU expects that IMT-Advanced would be developed as a new wireless access technology around the year 2012. The new network should be capable of supporting a bandwidth in the range of 20 MHz up to 100 MHz per carrier, with a high data rate up to 100 Mbps in a high mobility conditions. Such mobility conditions include vehicular (speeds up to 120 km/h) and high-speed vehicular (up to 350 km/h). IMT-Advanced will also provide a speed of 1 Gbps for low mobility such as stationary (fixed or nomadic terminals) and pedestrian (speeds up to 3 km/h) [4]. Seamless applicability with both mobile networks and IP networks (global roaming capabilities) is another feature of IMT-Advanced connectivity technology. Additionally, unicast, multicast broadcast services and multiple radio interfaces with the seamless handover technique will address the cellular network with a good coverage [20].

It is foreseen that IMT-Advanced systems shall be able to support Multiple Input-Multiple Output (MIMO) and beamforming. Thus, it should have the capability of supporting multi-antenna at the both receiver and transmitter ends to enhance the radio coverage [4].

Table 2.3 shows the IMT-Advanced deployment scenarios which require availability for mobile access for nomadic users, for ad hoc network users, for outdoor users (wide and metropolitan range) and for moving users (in a car or a high-speed train) [20].

**Table 2.3**  IMT-Advanced deployment scenarios

| Cell range | Performance target |
| --- | --- |
| Up to 100 m | Nomadic performance, up to 1 Gbit/s |
| Up to 5 km | Performance targets for at least 100 mbps |
| 5–30 km | Graceful degradation in system/edge spectrum efficiency |
| 30–100 km | System should be functional (thermal noise limited scenario) |

## 2.6   Interference Types and Assessment Methods

Interference can be classified into two categories: co-channel interference and adjacent channel interference [21]. Co-channel interference is defined as the interfering (unwanted) signal that has the same carrier frequency as the useful (wanted) information signal [22]. The adjacent channel interference (ACI) used in various CEPT ECC PT1 and ITU-R Working Party 8F studies on IMT-2000 are stated than the level of interference received, depending on the spectral leakage of the interferer's transmitter and the adjacent channel selectivity (blocking) performance of the receiver. Three possible interference problems have been identified in this chapter. Each of the problems is highlighted as follows:

1. Using same frequency by IMT-Advanced in the same geographical location will cause in-band interference to FSS earth station. However, signal received by FSS receiver is very weak due to the long distance between GSO and FSS earth station [5].
2. If IMT-Advanced operates in the adjacent channel to FSS channel, it will cause an ACI due to the out-of-band emissions. However, transmitting a terrestrial signal in the 3400–3600 MHz band can create interference in non-overlapping parts of the 3400–4200 MHz band.
3. LNB saturation in which the FSS receiving chain LNB becomes saturated if the incoming power is higher than −50 dBm. It will show a non-linear behaviour until it reaches the saturation region.

The first two interference types are the most interesting, as reported in this book. However, the LNA has utmost margin of interference limitation. For example, a classic LNA gain of −60 dBm gives a saturation level at the input of −50 dBm. Therefore in the worst case, the acceptable composite power at the input of a typical LNA would be around −55 dBm. This value has less impact on the scenarios proposed in this book, compared to the first two interference types [8].

In terms of interference assessment methods, different methods have been defined to assess the severity of interference in the European Conference of Postal and Telecommunications Administrations [23]. One of the methods was ISOP (Interference Scenario Occurrence Probability). It is the probability of having one terminal in the Interference Area (IA). In fact, the Interference Area is the domain of no acceptable interference which can accrue relative to the area of the cell or sector. This measure is related to the number of terminals deployed in a cell or sector and

**Fig. 2.5** Interference
protection criteria

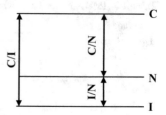

possibly to the cell planning methodology; it is used for BS to Subscriber Station (SS) interference between adjacent blocks in rural LOS situations as in [24].

Monte Carlo method has been used in Ref. [23] to evaluate the interference probabilities between terminals, since the placement of terminals or subscribers is rather random. The works reported in Refs. [15, 25–30] have used the Monte Carlo method to analyse intersystem interference between different systems, especially for mobile services. This method is based on statistical calculations of the number of interference cases in relevant to the deployment area.

Minimum coupling method (MCL) is based on system parameter calculations which can affect both transmitted and received power to ensure that the interference is below a given threshold in all cases [31]. In the scenario where the FSS earth station is affected by harmful interference, an assessment based on the IA is not so adequate since the whole cell could be blocked. Therefore, for the BS-BS interference, the worst case analysis is preferred.

It is not envisaged that a Monte Carlo analysis would give a very different results from the MCL scheme in the specific sharing situation as stated in Ref. [15]. The minimum separation distance can be calculated by adjusting the system parameters to estimate the interference according to the threshold level. This can be achieved by gradually increasing the separation distance until it meets the acceptable level for sharing and coexistence criterion.

## 2.7 Coexistence Criterion and Interference Model for Minimum Separation Distance

Intersystem interference is classified into two types. The first is the short-term interference which is a small percentage of the time in the range of 0.001–1.0% of the total received signal. The second type is the long-term interference in which the interference persists for 20% of the time. The first type is rarely evaluated in the coordination literature of the C-band, as it is much statistical in nature. Other reasons are that it is not often encountered, and it will be specific to the cases considered [32, 33]. Therefore, long-term interference has been considered in this book.

The interference protection criteria can be defined as an absolute interference power level $I$, interference-to-noise power $I/N$ ratio or carrier-to-interfering signal power $C/I$ ratio as described in Fig. 2.5. ITU-R F.758 has provided the details of two generally accepted values for the interference-to-thermal-noise ratio ($I/N$) for long-term interference into Fixed Service receivers. When considering interference from other services, it identifies $I/N$ value of –6 dB or –10 dB matched to the specific requirements of individual systems. The difference in decibels between carrier-to-noise ratio ($C/N$) and carrier-to-noise-plus-interference ratio ($C/(N + I)$) is known as a receiver sensitivity [32].

At the same time, an $I/N$ of $-10$ dB becomes the fundamental criterion for coexistence [6], so:

$$I - N \geq \alpha \tag{2.11}$$

where $I$ and $N$ are the interference level and noise floor of receiver, both in dBm, respectively, and $\alpha$ is the protection ratio in dB. If the $I/N = -10$ dB, then it implies that the interference must be 10 dB below thermal noise as shown in Fig. 2.5.

In order to determine the minimum acceptable level of the in-band interference signal into MEASAT FSS receiver, a minimum bandwidth should be identified. The 156.6 kHz channel bandwidth is considered as a case study for Internet access (see Appendix B for MEASAT-3 coverage and C-band specifications). The limit of the in-band-interference characteristics can be calculated as follows:

$$\frac{C}{I} = \frac{I}{N} + \frac{C}{N} = \left(10 + 5.7\right)\text{dB} \tag{2.12}$$

$$C = \frac{C}{N} + 10\log\left(KTB\right)\text{dBW} \tag{2.13}$$

$$I_{\text{inband}} = \left(C - 15.7\right)\text{dBw} = -166\text{dBW} \tag{2.14}$$

where $C$ (dB) is the carrier power at the receiver, $C/N$ (dB) is the required carrier to noise ratio, $I_{\text{in-band}}$ is the required protection ratio, $K$ is Boltzmann constant $= 1.38 \times 10^{-23}$ $J/K$, $T$ (K) is the temperature and $B$ is the noise bandwidth in Hz [33]. So, the maximum possible level of in-band interference is $-166$ dB (or $-136$ dBm) [34].

Increasing the noise floor level by a few dB can impact the existing licensed systems adversely. Though the subscribers will suffer service interruptions during the operating time of interferer systems, these interruptions are affected by a number of elements such as radio coverage, system capacity, reliability of data throughput and quality of voice service [35].

It is found in Refs. [3, 18, 36] that the interference avoidance measurements for fixed satellite earth station must detect space-to-earth transmission power much less than the thermal noise floor of the terrestrial receivers. This could increase their complexity, such as the interference from all other users causing not more than 0.4–0.5 dB [16, 21] degradation to the receiver threshold which is noted as receiver desensitization. Therefore, the proposed ($I/N = -10$) is seen as a prerequisite for a desensitization-proof receiver. This is justified by the deteriorating increase in the receiver noise floor by 0.4 dB.

The interference model depends on the spectral emission mask in which $I/N$ ratio are calculated after applying spectral emission mask [37]. This model can be used to determine both co-channel and adjacent channel interference. The model expression is given by [37]:

$$I(\Delta f) = P_t + G_t + G_r - A_{tt} \tag{2.15}$$

where $P_t$ is the transmitted power of the interferer in dBm, $A_{tt}$ is the attenuation due to the propagation in dB and $G_t$ and $G_r$ are the gains of the interferer transmitter and the victim receiver antennas, both in dBi, respectively. The thermal noise floor of victim receiver is given by [11]:

$$N = -114 + \text{NF} + 10\log_{10}\left(\text{BW}_{\text{victim}}\right) \tag{2.16}$$

where NF is noise figure of receiver in dB, and $\text{BW}_{\text{victim}}$ represents victim receiver bandwidth in MHz.

The protection distance is the distance necessary between the interfering transmitter antenna and the victim receiver antenna in order to protect the later from the harmful emissions. Thus the interference is either zero in an ideal case or minimum in real-life cases. The distance is usually calculated by using the MCL and an appropriate propagation model. The calculation of the protection distance is typically based on the characteristics of the source transmitter, the propagation channel and the recipient receiver.

The FSS off axis angle can affect the antenna receiving gain. Nevertheless, for a given off axis angle from the main receiving beam of the station, the victim might receive a different interference power. The $G_{\text{vs}}(\alpha)$ of a typical 1.8 m receiving FSS antenna is given by [5, 38]:

$$G_{\text{vs}}(\alpha) = G_{\max} - 2.5 \times 10^{-3}\left(\frac{D}{\lambda}\varphi\right)^2 \quad 0 < \alpha < \varphi_m \tag{2.17}$$

$$G_{\text{vs}}(\alpha) = 52 - 10\log\left(\frac{D}{\lambda}\right) - 25\log(\alpha) \quad 3.6° < \alpha < 48° \tag{2.18}$$

$$G_{\text{vs}} = -10\text{dBi} \quad 48° < \alpha < 180° \tag{2.19}$$

where $G_{\max}$ is the maximum antenna gain (38 dBi), $D = 1.8$ (satellite diameter) and $\lambda$ is the wave length in meter and $\varphi_m$; it is represented in this equation:

$$\varphi_m = \frac{20\lambda}{D}\sqrt{G_{\max} - 2 - 15\log\left(\frac{D}{\lambda}\right)} \tag{2.20}$$

The separation distance will become very large without mitigations because of the high sensitivity of FSS receiver as explained in Eq. (2.14). The shielding technique ($R$) can attenuate the interference power, where $R$ may take a value between 0 dB and 40 dB depending on the materials and shielding arrangement, as clarified below:

$$20\log(d) = -I + \text{EIRP}_{\text{Interferer}} - 92.5 - 20\log(F) - A_h + G_{vs}(\alpha) - R \quad (2.21)$$

where $A_h$ is the factor related to the territories as described earlier in this chapter, $d$ is the separation distance, $R$ is the shielding loss, EIRP is the effective isotropic radiated power transmitted from the interferer, $F$ is the frequency and $G_{vs}$ is related to the typical receiving FSS antenna gain [2, 39].

## 2.8  Spectrum Allocation and Harmonization

Frequency spectrum plan, otherwise known as spectrum allocation, is obtained from the regional or global spectrum fragmenting plan. The scale economy of each country has its own special needs, so, if these spectrum segmentations are unable to benefit these needs, the spectrum plan should be realigned to benefit the local economy. Therefore, innovative approaches should continue to pop up in order to benefit from the divergent assets. However, it should be noted that spectrum allocations include some of necessary recourses which cannot be moved.

The spectrum management cannot be replaced with the technological improvements. Even though, these improvements may need spectrum managers to allocate adjacent spectrum bands. These adjacent bands can be paired or used with spectrum fragments [40].

Harmonization is driving the need to develop alternative technology to increase spectrum usage and benefits. Consequently, not following a good harmonization will issue a fragmented status at certain bands. Harmonization is a very important procedure to prevent the national spectrum plan of being isolated of the global harmonization. Accordingly, it is considered as the main goal in the spectrum policy [40]. However, without harmonization the same product should be customized and redesigned for different countries according to its frequency plan. These products will require extra manufacturing costs in lower volumes. Therefore, aligning country spectrum plan with the global one will significantly improve the quality of wireless devices and correspondingly lower the manufacturing cost because of the increasing customized design volume [41]. To achieve the harmonization into the allocation plan, several models can be followed to define the least restrictive technical conditions. Some of these models are most applicable to develop technical conditions for the users in order to get access to the spectrum plan. These models are briefly discussed in the following subsections.

In this book, ACIR will be used to determine the coordination and also to calculate the guard band separation required for targeted systems in order to coexist in a close geographical location.

### 2.8.1  Traditional Compatibility and Sharing Analysis Method Using ACLR and ACS

Sharing and compatibility studies have used this model for many years. It aims to share same adjacent frequency bands among different services. This model requires referencing the knowledge of the characteristics of the new transmitted and received signal in order to be coordinated with the old one [42]. The adjacent channel leakage ratio (ACLR) and adjacent channel selectivity (ACS) are the most useful parameters in this model to investigate the adjacent frequency compatibility. However, combination of these two parameters will introduce the Adjacent Channel Interference Ratio (ACIR). Other important parameters in this model are bandwidth, ACLR, selectivity and blocking at the receiver side [43].

### 2.8.2  The Block Edge Mask (BEM) Approach

The Block Edge Mask (BEM) model that can be used for a point-to-multipoint fixed wireless system when a system is introduced without making decision on the operating frequency used by that system. The best method of spectrum access is provided by this model with a high flexibility. More than one frequency bands can be provided to the operator, in order to control the interference by changing the signal envelope and the transmitted power. Reference knowledge to the maximum acceptable in-block and out-of-block power should be available [43].

### 2.8.3  The Power Flux Density (PFD) Mask Model

The aggregate PFD is calculated by adding all the power flux densities of the transmitters around the susceptible receiver to identify the acceptable level of interference in order to give licence to neighbours. This model determines the maximum expected aggregated power received by the victim. This method is different from the BEM approach because the latter depends on the emission power of one interferer. After estimating the expected interference level, this model may allow the neighbouring services to operate with a consideration of several parameters to give flexibility for spectrum usages [42].

**Fig. 2.6** A sketch of a
hypothetical spectral
emission mask [44]

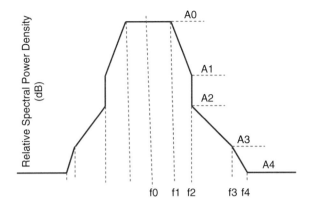

Additional complications may accrue using this model due to the assumption of deployment density for different geographical locations. This assumption should be based on a good understanding to the extensive deployment of current services. However, either an increasing or decreasing network deployment density will have an impact the spectrum allocation. Definitely, careful monitoring of the future services deployment will disapprove this model for long-term deployments.

### 2.8.4   Hybrid Model

It is based on a combination of the two models described in Sects. 2.8.2 and 2.8.3. It is used to allocate the frequencies of different systems with more flexibility [42]. Based on the previous models, it is concluded that accountability and responsibility of choosing the proper model according to the type of service is important to achieve the spectrum harmonization. Flexibility can be achieved by using the traditional analysis of compatibility and sharing models based on several factors. Technical characteristics represent the most important factor among them. However, the main limitation of these models is the technology-specific input parameters, whereby any change of technology could invalidate the results.

## 2.9   Spectrum Emission Mask

The main task in coexistence assessment is to compile relevant transmitter and receiver characteristics as well as parameters of each system to be modelled for intersystem interference estimation. The main interference factor for the IMT-Advanced BS is an interferer other than antenna gain, and the transmitted power is the Spectrum Emission Mask (SEM). However, SEM is a graphical illustration powered by the rules set by regulatory bodies such as FCC and the European

Telecommunications Standards Institute (ETSI) [44]. Figure 2.6 shows that a typical mask has a piecewise linear shape.

The spectral emission mask is divided into a number of segments with power spectral density attenuation on y-axis and frequency spacing on x-axis.

SEM can be used to generate the worst case of interference from the interferer side. Therefore, the coexistence study can be applied using SEM as a critical parameter. This fact should be identified by regulators for a safe coexistence in the co-channel and adjacent channel. A guard band, specified according to the SEM, can be used between the interfering systems operating in adjacent regions to protect the victim receiver from any out-of-band interference.

## 2.10    Antenna Shielding Mitigation Technique

Shielding is used to attenuate the Electromagnetic Incompatibility (EMI) between sources (IMT-Advanced) and susceptible equipment (FSS receiver). The mechanism of shielding is described as follows: when terrestrial waves hit the shield, a part of its energy will be reflected because of the shield surface; another part of the energy will be absorbed and transformed to other shapes of energy (either thermal or electrical energy). Part of the electrical energy will be discharged through the ground, and the rest will pass through the shielding. So, basically the site shielding is about physical obstruction built to reduce the interference from the interferer to the victim receiver [45].

Measurements for FSS have been discussed in details, especially the shielding of FSS at 3400–4200 MHz [45–47]. In addition, most of the studies recommended that shielding can reduce the harmful interference [38, 48–53]. The best isolation may happen when the enclosure is fabricated as one homogeneous piece. A small opening may be designed on the shielding for maintenance, repair or system upgrade. Same material should be considered for the opening, which is like cover, door or window. The shielding material choice is wide, but each material differs in its ability to attenuate the electromagnetic waves.

Shielding can be natural by locating the FSS dish in around the back of building or hill. It can be done artificially by adding one or two walls on the path between victim and interferer [45]. Two walls will be much better because it will duplicate the amount of attenuation. The dish elevation angle should be considered during shielding deployments to prevent signal blocking by the shielding shape [46]. By putting the dish as low as possible, and with high shielding all around except in the direction of the beam to satellite, will help to avoid the interference. The signal reception will be better, because the more the dish is hidden the greater the reduction in interference level [47].

**Fig. 2.7**  Switched-beam
antenna configuration [58]

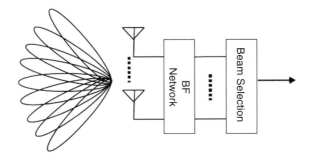

## 2.11  Mitigation Using Smart Antenna Technologies

The antenna associated with each radio transmitter receives the system output power
and radiates it as radio waves [54]. Smart antenna is an issue of the tested antenna
beam discrimination. However, antenna discrimination is defined as the differential
gain compared to the maximum gain for an antenna in the specified direction.
Usually, masks are provided for the main lobe, the first side lobe and other side
lobes [55]. By using smart antenna for nulling the interference power sent to the
victim, power discrimination can be achieved.

Conventional discrimination loss is caused by using different antenna polarization and antenna directivity misalignment of both the interferer transmitter and victim receiver antennas. However in the smart antenna system, the power discrimination
can be achieved by synthesizing the main lobe to the user. This is done by nulling
the side lobes in order to mitigate the power in the direction of victim. Antenna
direction misalignment between the interferer transmitter and victim receiver services is dependent on the off axis angles between the victim and interferer. For any
off axis angles to the victim receiver, the interference effects will be smaller [56,
57].

Smart antenna systems, by using spatially separated antennas, referred to as
antenna array, maximize the signal-to-interference-plus-noise ratio (SINR) of the
received signals and suppress interferences and noise power by digital signal processing after analogue to digital conversion. Smart antennas can use spatial domain
processing by using multiple antennas, thus enabling them to have the intelligence
to process the data at both receiver and transmitter.

The smart antennas are often classified as switched-beam arrays and adaptive
array antennas [58, 59]. The switched-beam arrays comprise beamforming networks and a beam selection processor, as shown in Fig. 2.7. The processor selects
the beam with maximum power controlled by switching the beams. However, the
adaptive array antennas incorporate more intelligence than the switched-beam
arrays.

Adaptive antenna arrays can estimate their environment in accordance with the
propagation channel between the receiver and the transmitter. This information is
then used to weigh the data received at transmitter from the antenna array to

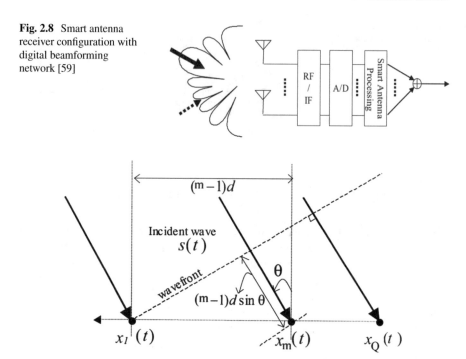

**Fig. 2.8** Smart antenna receiver configuration with digital beamforming network [59]

**Fig. 2.9** Path difference in planar wave model (linear array of *Q*-element)

maximize the response for the desired user. Figure 2.8 illustrates the typical configuration of the adaptive array antennas.

The processor determines the optimum weight vector for a given environment, and, thus, the adaptive antennas can combine the received signals, thereby maximizing the SINR and not merely SNR. In this book, the term smart antenna is used to describe the adaptive (array) antenna as in Ref. [60]. A smart antenna is therefore an intelligent antenna system different from a conventional omni-directional or fixed-beam antenna system, which merely receives and transmits signals without any adaptive behaviour to any change in environment [61, 62]. In order to explain the principle of smart antenna signal processing, the simple discrete wavefront model with narrowband signal is illustrated in Fig. 2.9.

It is assumed to have a uniformly and equally spaced linear array of identical and omni-directional *Q*-elements as array geometry. Let the angle between array normal and incident wave be $\theta$; the far-field expression of the electrical signal at the *m*-th element at any time t is given by [63]:

$$x_m(t) = S(t) \cdot \exp\left(-j\frac{2\pi}{\lambda}d(m-1)\sin\theta\right) + n_m(t) \dots m = 1, 2 \dots, Q \quad (2.22)$$

where $S(t)$, $\lambda$ and $\theta$ are the envelop, wavelength and direction-of-arrival (DOA) of an incident wave, respectively, $n_m(t)$ is the additive white Gaussian noise (AWGN) at the $m$-th element and $d$ is the space between each antenna. In Eq. (2.22), if $S(t)$ is a narrowband signal, the temporal delay caused by path difference between the elements corresponds to the phase difference. The output of the antenna array is produced by the inner product (multiply-accumulate operation) of input signals and weight coefficients determined by adaptive algorithms as:

$$y(t) = \sum_{m=1}^{Q} V_m^* x_m (t) = V_1^* x_1 (t) + \ldots + V_Q^* x_Q (t) \tag{2.23}$$

where $V$ is the weight and $V^*$ is the channel weight representation. Equation (2.23) can be also rewritten by vector expression as:

$$y(t) = \mathbf{V}^H \mathbf{x}(t) \tag{2.24}$$

The transmitted power can be estimated using Eq. (2.24):

$$\begin{aligned} P(w) = E(P(t)) = E\left(\left|y(t)\right|^2\right) &= E\left[\mathbf{v}^H \mathbf{x}(t) \mathbf{v} \mathbf{x}(t)^H\right] \\ &= \mathbf{v}^H \mathbf{v} E\left(\mathbf{x}(t)\mathbf{x}(t)^H\right) \\ &= \mathbf{V}^H \mathbf{R}_{xx} \mathbf{V} \end{aligned} \tag{2.25}$$

where $\mathbf{R}_{xx}$ is the autocorrelation of signal $\mathbf{x}$ and where the superscript $H$ denotes Hemitian transpose operator. The data vector $\mathbf{x}$ is written by:

$$\mathbf{x}(t) = \mathbf{a}(\theta) s(t) + \mathbf{n}(t) \tag{2.26}$$

where $\mathbf{a}(\theta)$ denotes array mode vector or eigenvalue (see Appendix C for mathematical expression and notifications) as:

$$\mathbf{a}(\theta) = \left[1,,,e^{-j2\Pi\frac{d}{\lambda}\sin\theta},,,\ldots,,,e^{-j2\Pi(Q-1)\frac{d}{\lambda}\sin\theta}\right]^T \tag{2.27}$$

In the absence of interference, the adaptive beamforming has the maximum SNR. In the presence of interferences, the SINR will be maximized by reducing the power of the interference. If a single planar wave arrives through a single path, Eqs. (2.24), (2.25), (2.26) and (2.27) show that the optimum weight vector becomes the mode vector of the incident wave [64].

Several techniques have been proposed to determine the optimum weight, such as minimizing mean square error (MMSE), which the solutions are based on solving Wiener-Hopf equation, maximizing signal-to-noise ratio (MSN) [65]. This technique is based on generalized eigenvalue problem and linearly constrained minimum variance filter (LCMV) or directional constrained minimum power (DCMP).

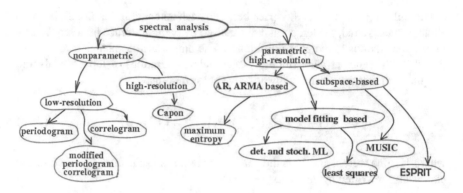

**Fig. 2.10** A handful of existing approaches for spectral estimation

This technique requires the DOAs information of the desired signals and interferers as a priori. It requires some channel information, such as the training sequence in the case of MMSE and the directional information of incident signals in the case of MSN and LCMV. Besides these methods, there are also blind methods, such as constant modulus algorithm (CMA), in which the directional information is not necessary. However, the spatial channel signature such as DOAs of incident signals is often required for efficient spatial domain processing.

Therefore, adaptive beamforming algorithms are classified as DOA-based and temporal reference based. The proposed method is based on the DOA algorithm passing the DOA information to the beamformer. Temporal reference beamformers can use a known training sequence in order to adjust the weights to form the pattern towards the SOI with the nulls in the FSS_ES direction.

In this book the least mean square (LMS) algorithm will be used as a beamforming algorithm, though it is commonly used to adapt the weights. It is a low-complexity algorithm that requires no direct matrix inversion and no memory.

## 2.11.1   Direction-of-Arrival Estimation

Theoretically, the maximum $(Q - 1)$ interferers can be cancelled with the M-element array. According to the literature, DOA is divided into parametric and non-parametric. For the parametric methods, they are all high resolution such as maximum entropy [66], maximum likelihood sequence detection [67] and subspace methods [68, 69].

Regarding the non-parametric spectrum analysis, it is usually based on high-resolution or low-resolution methods. The high resolution could be like capon spectrum analysis [70], low resolution like: periodogram normal convergence properties of correlogram spectral estimates [71], spectral analysis of nonuniformly sampled data using the periodogram [72] and modified periodogram averages [73]. Figure 2.10 shows several types of DOA estimation.

These types of DOA techniques are based on the angle domain approach that exploits the fading correlation among antenna elements. Tracking techniques can provide a statistical average performance of DOA estimation in a fading environment with small array spacing [74]. The main features of the angle-domain approach using DOA estimations are as follows:

1. Application for urban cellular base station with small angle spread [75].
2. No straight diversity gain; however, angular diversity can be available by combining multipaths [74].
3. Large computational load is required for DOA estimation [75].

The basic method in DOA estimation algorithms with an antenna array is the beamformer method. This method is based on the same principle as the Fourier transforms [76, 77]. Other methods, such as Capon method, involve measurement of electrical power by scanning all the directions while keeping a null radiation pattern to other arrival waves using the antenna array [78]. The resolution of the Capon method depends on the beamwidth of the main beam in order to scan the main beam at all angles. A narrow beamwidth is necessary in order to achieve high resolution. This can be done by using a large numbers of elements.

On the other hand, it is possible to estimate DOA with high resolution by scanning all angles with a null. This is because the beamwidth of the null is narrower than the main beam. Based on these considerations, the linear prediction (LP) method is proposed in Ref. [79]. Moreover, the Min-Norm method [80] and Pisarenko method [81] have also been proposed. Weight assumptions of these methods are extended from the LP method. Although DOA estimation algorithm mechanisms have been developed and added to the adaptive array antenna, DOA estimation algorithms use the characteristics of the adaptive array antenna, because its principle is closely related to the adaptive array antenna [82, 83]. For example, the principle of the Capon method is equal to the DCMP adaptive array antenna [84], and the LP method is equivalent to the side lobe canceller and a power inversion adaptive array antenna [85, 86].

To resolve the problems of accurate DOA, the multiple signal classification (MUSIC) method was proposed [87]. The MUSIC method makes use of the eigenvectors correlated to the noise subspace of a correlation matrix which are orthogonal to the mode vectors that express phase differences to a base point. Although the computational method of the MUSIC method is more complex than the beamformer method, the MUSIC method has higher resolution than conventional methods [88, 89]. The MUSIC method cannot estimate DOA correctly when the accuracy of the correlation matrix is insufficient, and it has higher computational costs due to scanning the MUSIC spectrums [68].

Due to these problems, the Root-MUSIC method has been proposed [90]. Moreover, the Unitary MUSIC method which uses the unitary transformation that can replace the calculation of the complex numbers with real numbers. This method was proposed in order to reduce the computational complexity and to achieve better speed of the algorithm [91, 92]. On the other hand, the estimation of the signal parameters via rotational invariance techniques (ESPRIT) is used to compute phase differences between subarray antennas [93].

A spectrum search is unnecessary in ESPRIT, although it has essentially the same accuracy as the MUSIC method. ESPRIT can be classified into LS-ESPRIT (Least-Squares) [93] and TLS-ESPRIT (Total-Least-Squares) [94] according to the methods used for calculating the phase differences of two subarrays. The unitary transformation method has been proposed as well as the case of the MUSIC method [95, 96]. Moreover, these algorithms have been extended to 2D or 3D. In the same time, it can simultaneously estimate not only DOA in the horizontal plane but also the directions of elevations and time delay [97–100]. These methods are characterized by successively updating the eigenvectors in the signal subspace of a correlation matrix.

## 2.11.2   Direction of Arrive Estimation Using MUSIC Algorithm

The MUSIC algorithm is a type of DOA estimation technique based on eigenvalue decomposition (EVD), and it is also called a subspace-based method [68]. It has many advantages including simpler implementation and higher resolution than any other subspace-based method.

For MUSIC data modelling, if the basic model of the narrowband signal $S_i(t)$ for the $i$-th source, where $i = 1, 2 ..., L$, the signals received at the M-element antenna array spaced by a half wavelength can be modelled as [68]:

$$\mathbf{X}(t) = \mathbf{L}s(t) + \mathbf{n}(t) \tag{2.28}$$

where the array output $\mathbf{X}(t)$ is a snapshot vector, $s(t)$ is the signal and $\mathbf{n}(t)$ is the complex AWGN vectors at time $t$. The columns of the channel matrix $\mathbf{L} = [L_1, L_2, ...,$ and $L_m]$ consist of the spatial channel vectors for $P$ sources. The spatial channel vector $L_i$ for the $i$-th source can be provided by the array response vector $a_i$ under the assumption that the plane waves arrive at an ideal omni-directional antenna array from the point sources as:

$$L_i = a(\theta_i) = \left[ 1,,,e^{-j\Pi \sin \theta_i},,,......,,,e^{-j\Pi(Q-1)\sin \theta_i} \right]^T \tag{2.29}$$

where $\theta_i$ is the DOA for the $i$-th source, and the superscript $T$ denotes the transpose operator.

The correlation matrix of $x(t)$ is given by:

$$\mathbf{R}_{xx} = E\left[ x(t)x^H(t) \right] = \mathbf{V}\mathbf{R}_{ss}\mathbf{V}^H + \sigma_w^2 \mathbf{I}_Q \tag{2.30}$$

where $E$ and the superscript $H$ denote the statistical expectation and Hermitian conjugate operators, respectively. $\mathbf{R}_{ss} = E[s(t)s^H(t)]$ is the signal covariance matrix, and

$\sigma^2$ represents the noise variance. Since the correlation matrix $R_{xx}$ is a positive definite Hermitian, it can be decomposed to signal and noise subspaces by the complex-valued EVD as:

$$\mathbf{R}_{xx} = \mathbf{U}\varLambda\mathbf{U}^H \tag{2.31}$$

where $\mathbf{U}$ is a unitary matrix composed of eigenvectors, and $\varLambda$ is diagonal $\{\lambda_1, \lambda_2,...\lambda_Q\}$ of real eigenvalues ordered by $\lambda_1 \geq \lambda_2 \geq ... \geq \lambda_Q > 0$. If an eigenvector $e_i$ is orthogonal to $\mathbf{V}^H$ of rank $L$,
   then $e_i$ is an eigenvector of $\mathbf{R}_{xx}$ with the eigenvalue of $\sigma^2$ as:

$$\mathbf{R}_{xx}e_i = \left(\mathbf{V}\mathbf{R}_{ss}\mathbf{V}^H + \sigma^2\mathbf{I}\right)e_i = \sigma^2 e_i \tag{2.32}$$

The eigenvectors of $R_{xx}$ with eigenvalue of $\sigma^2$ lie in the null space of $\mathbf{V}^H$. On the other hand, some eigenvectors lie in the range of $\mathbf{V}$. They can be partitioned into signal and noise components. In a similar manner, the correlation matrix can be also partitioned as:

$$\mathbf{R}_{xx} = \mathbf{U}_s\varLambda_s\mathbf{U}_s^H + \mathbf{U}_n\varLambda_n\mathbf{U}_n^H \tag{2.33}$$

where $\mathbf{U}_s$ and $\mathbf{U}_n$ are the unitary matrices of signal subspace and noise subspace, respectively, while $\varLambda_s$ and $\varLambda_n$ are the diagonal matrices of the eigenvalues.
   The noise subspace eigenvectors of corresponding eigenvalues of $\sigma^2$ are orthogonal to the signal subspace and eventually, orthogonal to the array response vectors. Based on this, the MUSIC spectrum is typically expressed by [68]:

$$\mathrm{PMU}(\theta) = \frac{\mathbf{a}^H(\theta)\mathbf{a}(\theta)}{\mathbf{a}^H(\theta)\mathbf{E}_n\mathbf{E}_n^H\mathbf{a}(\theta)} \tag{2.34}$$

where $E_n$ is the matrix that has columns consist of noise subspace eigenvectors. Using Eq. (2.34), the peaks appear at the DOAs of incident signals [101].

## 2.11.3   High-Resolution DOA for Uniform Linear Array (ULA) Using Improved-MUSIC Algorithm

The first serious discussions and analyses of subspace-based DOA emerged during the 1986s with the first proposal of Multi Signal Classification (MUSIC) algorithm [68], which uses the noise-subspace eigenvectors of the data correlation matrix to form a null spectrum and yield the corresponding signal parameter estimation. Mathematical modelling and analysis for the MUSIC power spectrum have been investigated with different signal-to-noise ratio (SNR), and pilot coherent to a largest local minimum values and the source parameter has estimated as in Ref. [102].

**Fig. 2.11** Calculating the
angle of arrived signal

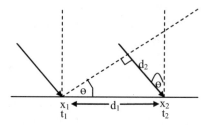

Recently, researchers have given attention to develop the resolution of MUSIC algorithm which has an attractive use in a critical mission of wireless services [103–105]. However, a precise DOA estimation is obtained by MUSIC algorithm compared to maximum likelihood method and estimation of signal parameters via rotational invariance techniques (ESPRIT) algorithm, which is used in order to separate and estimate the phase shifts due to delay and direction-of-incidence [106].

High-order cumulative methods require the signal statistical properties, and it needs larger snapshots to get a good performance. Besides that, it has a heavier computation load. Alternatively, it has been proved that MUSIC-Like can exploit the cyclostationarity in order to increase the resolution power and noise robustness. However, both cyclic and conjugate cyclic correlation matrices are used so that it introduces more complexity and applicable with limited conditions [107].

The Conjugate Augmented MUSIC (CAM) algorithm mentioned in Ref. [108] is a second-order statistical approach of the received signals to get the conjugate steering matrix, together with steering matrix. CAM is better than MUSIC-Like algorithm in terms of number of directions, estimation capacity, angle resolution, required snapshots and immunity to noise. In Ref. [109] multi-invariance MUSIC (MI-MUSIC) algorithm has been introduced in order to prove that it has a better angle and delay estimation performance than ESPRIT and MUSIC algorithm.

### 2.11.4   The Angle of Arrived Signals Calculation

As mentioned in the literature, the angle of arrival can be identified by multiplying the difference in time between two signals. Thus, the obtained angle of arrival is depending on the first received signal. This is illustrated in Fig. 2.11.

Figure 2.11 shows that two signals have arrived with a specific time delay. It utilizes the value of the time delay with the light speed which can obtain the delayed eigenvector ($d_2$ in Fig. 2.11), as given by:

$$\tau = t_2 - t_1 \tag{2.35}$$

where $t_1$ is the arrival time of the first signal, $t_2$ is the arrival time of the second signal and $\tau$ is the time difference between first and second signals. From the essential formulas of velocity, the eigenvalue can be calculated as:

$$d_2 = \upsilon \times \tau \qquad (2.36)$$

where $d_2$ is the value of the delayed vector in meter and $\upsilon$ is the velocity in meter per second. Accordingly, the $d_2$ can be calculated as:

$$d_2 = d_1 (x_2 - x_1) \sin \theta \qquad (2.37)$$

$$\theta = \sin^{-1} \left( \frac{d_2}{d_1 (x_2 - x_1)} \right) \qquad (2.38)$$

where $\theta$ is the angle between the imaginary perpendicular line and the second vector. By using $90-\theta$, the direction of the arrived signal can be identified.

## 2.12  Previous Assessments of Interference Between FSS and Broadband Terrestrial Services

The coexistence study of Asia Pacific Telecommunication (APT) has reported a harmful interference received by the FSS because of BWA transmitter. The reported cases cover interference for BWA in overlapping frequency bands and in non-overlapping bands. A practical case study has observed same interference at one location in Indonesia and Pakistan as provided in Ref. [49]. Another case study provides the results of a set of experiments carried out on the impact of interference into a commercial TVRO terminal in the 3400–4200 MHz band in Japan. In this study, the interference is caused by terrestrial station, and the subsequent measurements were performed using the traditional filtration technique in order to avoid the ACI [49].

The effect of different FSS receiver elevation angles has been analysed in Ref. [49]. The result of report [49] shows that sharing is impossible within the radius of 0.5 km. In this case, the measured BWA signal is 30 dBm, whereas the measured FSS signal is −146.5 dBm. In addition, rigorous calculations for the different elevation angle effect were done in ITU-R SF.1486, ITU-R S.1432, ITU-R P.452 and ITU-R BO.1213 propagation models [2, 11, 48, 110]. Other techniques that produce almost similar results for the coexistence between terrestrial communication and FSS receiver in the 3400–4200 MHz band have been discussed in Refs. [50, 111–113].

Spectrum sharing and most possible flexibility have been discussed in Ref. [114], and it is concluded that mutual power interference cannot be solved without considering the SEM (which is an issue of physical layer). At the same time, dynamic interference avoidance technique is inapplicable in the case of terrestrial effect on FSS receivers because FSS is designed as a receiver only. A new approach has been proposed to solve the interference based on Monte Carlo techniques in Ref. [114] which is still not feasible because it produces a large separation distance.

Measurement of pulsed co-channel interference due to unwanted emission in the 4 GHz fixed satellite earth station has been verified in Ref. [115]. A pulsed signal at the same frequency was sent to the victim receiver through the front end of FSS unit, which functions as a television receives-only (TVRO). Interference was tested for the lower pulse rates and widths, and due to these cases, the results showed that interference may increase by 50 dB above the carrier level, which is an indication of a serious signal interruption [115].

The Inter-American Telecommunication has concluded that co-channel operation of FSS earth stations and BWA systems will have severe constraints on both FSS and BWA [116].

WiMAX forum suggested that co-frequency is possible between WiMAX and FSS using site engineering. Each case of deployment should be treated separately using some mitigation techniques to coordinate this service with other services, which have known locations. If the WiMAX system is deployed at every location, then mitigation techniques will be necessary, to avoid the interference to other systems like FSS. It will be more difficult to coexist with the two services, even at a sub-bands, if the satellite earth station is using a wideband front-end RF. These wideband filters cannot cut off the WiMAX out-of-band emission [45].

In January 2006, the Office of the Telecommunications Authority (OFTA) of Hong Kong established measurement experiments to test the interference from BWA, which operate at 3.4–3.6 GHz frequency band, to the fixed satellite services, which operate at 3.4–4.2 GHz frequency band. The FSS used to receive a television signal via 3 m antenna diameter. The FSS received TV signal at 3725 MHz with a bandwidth of 6 MHz from the AsiaSat-3S satellite at a subsatellite longitude of 105.5° east (elevation angle 63°). Whereas BWA transmitted a signal at 3.55 GHz, with a 3.5 MHz channel bandwidth, a separation distance used was 360 m with a 12 dB EIRP and without down tilting. The study covered the down-tilting analysis, antenna gain and transmitting EIRP. When the down-tilting technique was used, the TV picture was frozen at 10°. But when EIRP reduced gradually to 9 dB, the picture has returned back to the TV. A 10 dB band-pass filter on the FSS front end could solve the problem of 10° down-tilting without using the EIRP reduction [46]. The case of the aggregate effect of multiple BWA base stations and terminals was not considered.

Sharing studies based on simulation have been conducted in Japan for the frequency of 3400–4200 MHz. The results of sharing studies based on the interference model using the existing ITU-R Recommendations as well as those of new sharing studies taking into account the shielding effect by the artificial objects observed in a real environment are also included. However, interference area ratio exceeds 10%, and the minimum separation distances were 76 km and 45 km, respectively, for 5° and 48° FSS elevation angle in rural area macro cell size. The best separation distances were 60 km and 8 Km, respectively, for 50 and 480 FSS elevation angle in urban area. Details of the experimental evaluation on the robustness of a TVRO terminal against interference from an IMT-Advanced transmitter in the 3400–4200 MHz band are explained in Ref. [47].

In Korea, the impact of single and aggregate cases of IMT-Advanced systems on an FSS earth station have been considered. The study also used the propagation model of Rec. ITU-R P.452 in order to evaluate the effect of terrain profile on the separation distance of the two systems (for frequency sharing). For a single base station of IMT-Advanced system without any terrain effects, separation distance between the BS of IMT-Advanced and FSS ES for co-frequency sharing is required from 39 km to 127 km. These distances are for the cases of the output power 47 and 53 dBm/72 MHz of BS and 48° and 5° of elevation angle of FSS ES, respectively. When the output power of a BS is 7 dBm/50 MHz and the elevation angle of FSS ES is 20°, the calculated separation distance without terrain effect is 62 km. Within this distance, the interference from the BS of IMT-Advanced system into FSS earth station would be more than 6%. When terrain effect was considered in this case, only small area within the radius of 62 km is subject to interfered with the 6% of criterion. For aggregate base stations of IMT-Advanced system, the required separation distances are varied from 43 to 164 km depending on the transmit power of IMT-Advanced BSs and FSS ES elevation angle. In case of aggregate mobile stations, required distances are from 0.5 to 1.5 km depending on user density [117].

In the WINNER research group [52, 118–121], the study obtained methods that could be envisaged based on their assessments of frequency sharing between IMT-Advanced and FSS. Generally, the base stations have tri-sectorial antennas, which are ways of reducing the transmitting output power level. This can disable the antenna sector in those points towards the FSS earth station, and more sectors will decrease the exclusion area. However, the base stations and terminals located in an FSS protection area could be monitored to ensure the usage of other frequencies.

In Korea a complicated study based on deterministic analysis and special simulation to improve sharing between FSS and IMT-Advanced considering Multiple Input-Multiple Output (MIMO) Special Division Multiple Access (SDMA) mitigation technique simulation has been investigated. This technique is used to mitigate the interference from IMT-Advanced in CCI and ACI. Though the technique is interesting, the results were corrupted with a large amount of errors. Minimum separation distance reduction obtained from the MCL method is 24% [47].

Other useful techniques include frequency segmentation between FSS earth stations and IMT-Advanced systems and finding sub-bands unused by FSS in particular geographical locations. This would allow operation of IMT-Advanced without a protection area in the vicinity of FSS earth station [117]. Moreover, the Russian researchers had issued a research paper through the ITU-R which concluded the coverage zone of IMT-Advanced with a different I/N criteria [53].

## 2.13   Summary

Frequency band between 3400 and 4200 MHz is the most promising frequency for future IMT-Advanced because it has sufficient bandwidth, higher antenna gain and minimum cost. Some sharing studies between FSS receiving earth stations and IMT systems indicate that the required separation distance becomes approximately several tens of kilometres, without considering the actual terrain propagation conditions. However, other studies have shown that if the actual terrain propagation conditions, such as the influence of artificial objects, are taken into consideration, the required separation distance is significantly reduced, and the interfered area becomes spatially limited. This can elucidate the increased possibility of the sharing between FSS and IMT systems. It is concluded that a new practical shielding mitigation technique is needed to achieve the minimum separation distance. This technique can further increase the possibility of sharing between these systems using guard band insertion between the two services. It is also concluded that antenna power discrimination is an important factor to achieve the feasible coexistence. Therefore, one possible method is to apply smart antennas on IMT systems in order to null the EIRP in the direction of the interfered FSS earth station.

## References

1. L.F. Abdulrazak, Z.A. Shamsan, T.A. Rahman, Potential penalty distance between FSS receiver and FWA for Malaysia. Int. J. Publ. WSEAS Trans. Commun.. ISSN: 1109-2742 **7**(6), 637–646 (2008)
2. ITU-R SF.1486, *Sharing Methodology Between Fixed Wireless Access Systems in the Fixed Service and Very Small Aperture Terminals in the Fixed-Satellite Service in the 3400–3700 MHz Band. ITU-R R WP4-9S, Geneva,* Switzerland, November 2000
3. ITU-R Recommendation F.1402, *Frequency Sharing Criteria Between a Land Mobile Wireless Access System and a Fixed Wireless Access System Using the Same Equipment Type as the Mobile Wireless Access System*, Geneva, Switzerland, May 1999
4. ITU-R Recommendation M.1645, *Framework and Overall Objectives of the Future Development of IMT-2000 and Systems Beyond IMT-2000*, Geneva, Switzerland, Jan 1999
5. Recommendation ITU-R P.452-12, *Prediction Procedure for the Evaluation of Microwave Interference Between Stations on the Surface of the Earth at Frequencies Above About 0.7 GHz*, Geneva, Switzerland, May 2007
6. D. Laster, J.H. Reed, Interference rejection in digital wireless communications. IEEE Commun. Mag **14**, 37–62 (1997)
7. Report ITU-R M. [LMS.CHAR.BWA], *Characteristics of Broadband Wireless Access Systems Operating in the Mobile Service for Frequency Sharing and Interference Analyses*, October 2005, http://datatracker.ietf.org/documents/LIAISON/file244.doc
8. Document AWF-3/17, *Assessment of Potential Interference Between Broadband Wireless Access (BWA) in 3.4–3.6 GHz Band and Fixed Satellite Service (FSS) in 3.4–4.2 GHz Band.* Office of the Telecommunications Authority (OFTA) Hong Kong, September 2006. http://www.esoa.net/v2/docs/public_cband/ESOA_CBand_APTReport.pdf
9. M.D. Yacoub, *Foundations of Mobile Radio Engineering* (CRC Press, New York, 1993), pp. 67–70
10. M.P.M. Hall, L.W. Barclay, *Radiowave Propagation*, vol 30 (IEE Publication, 1989), p. 296

11. Recommendation ITU-R P.452-9, *Prediction Procedure for the Evaluation of Microwave Interference Between Stations on the Surface of the Earth at Frequencies Above About 0.7 GHz*, Geneva, Switzerland, July 2003
12. R.A. Mosbacher, *U. S. Spectrum Management Policy: Agenda for the Future* (DIANE Publishing, 1994)
13. IST-2003-507581 WINNER, *WINNER Spectrum Aspects: Methods for Efficient Sharing, Flexible Spectrum Use and Coexistence*. D6.1 v1.0, (2004)
14. D.L. Schilling, L.B. Milstein, R.L. Pickholtz, F. Bruno, E. Kanterakis, M. Kullback, V. Erceg, W. Biederman, D. Fishman, D. Broadband, CDMA for personal communications systems. IEEE Commun. Mag **29**(11), 86–93 (1991)
15. L.F. Abdulrazak, K.F. Al-Tabatabaie, Preliminary design of iraqi spectrum management software (ISMS). Int. J. Adv. Res. **5**(2), 2560–2568 (2017)
16. CEPT ECC Rep. 100, *Compatibility Studies in the Band 3400–3800 MHz Between Broadband Wireless Access (BWA) Systems and Other Services*, Bern, Switzerland, Feb 2007
17. H.G. John, *Rural America at the Crossroads: Networking for the Future* (University of Maine, Augusta, 1991), pp. 74–86
18. R. Conte, Satellite rural communications: Telephony and narrowband networks. Int. J. Satell. Commun. Netw Wiley interScience Publisher. **23**, 307–321(2005)
19. M. G'erard, *VSAT Networks*, 2nd edn. (Wiley, USA, New York, 2003), pp. 48–77
20. A. Mihovska, R. Prasad, Secure personal networks for IMT-advanced connectivity. Springer Publ. Wirel, Pers. Commun. **45**(4), 445–463 (2008) ISSN: 0929-6212
21. IEEE Std 802.16.2-2004, *IEEE Recommended Practice for Local and Metropolitan Area Networks Coexistence of Fixed Broadband Wireless Access Systems* (IEEE, USA, New York, 2004)
22. P. Stavroulakis, *Interference Analysis and Reduction for Wireless Systems* (Artech House INC, UK, London, 2003), pp. 214–221
23. CEPT ERC Rep. 99, The analysis of the coexistence of two FWA cells in the 24.5–26.5 GHz and 27.5–29.5 GHz bands. *European Radiocommunications Committee within the European Conference of Postal and Telecommunications Administrations*, Edinburgh, (2000)
24. CEPT ECC Rep. 33, The analysis of the coexistence of FWA cells in the 3.4–3.8 Ghz band. *Electronic Communications Committee Within the European Conference of Postal and Telecommunications Administrations*, (2003), Available: http://www.erodocdb.dk/Docs/doc98/official/pdf/REC0506.PDF
25. D.L. Schilling, S. Ghassemzadeh, K. Parsa, Z. Hadad, Broadband CDMA overlay. Int. J. Wireless Inf. Networks **2**(4), 197–221 (1995)
26. CEPT ERC Rep. 64, Frequency sharing between UMTS and existing fixed services. *European Radiocommunications Committee within the European Conference of Postal and Telecommunications Administrations*, Menton, (1999)
27. ITU-R Rep. M.2030, Coexistence between IMT-2000 TDD and FDD radio interface technologies within the frequency range 2500–2690 MHz operating in adjacent bands and in the same geographical area. *International Telecommunications Union Radiocommunication-Sector Report*, (2003)
28. L.F. Abdulrazak, K.F. Al-Tabatabaie, Broad-spectrum model for sharing analysis between IMT-advanced systems and FSS receiver. J. Electron. Commun. Eng. (IOSR-JECE) **12**(1.), Ver. III), 52–56 (2017)
29. CEPT ECC Rep. 045, Sharing and adjacent band compatibility between UMTS/IMT-2000 in the band 2500–2690 MHz and other services. *Electronic Communications Committee within the European Conference of Postal and Telecommunications Administrations*, (2004)
30. ITU-R Rep. M.2113, Report on sharing studies in the 2500–2690 MHz band between IMT-2000 and fixed broadband wireless access systems including Nomadic applications in the same geographical area. *International Telecommunications Union Radiocommunication-Sector Report*, (2007)
31. CEPT ERC Rep. 101, A comparison of the minimum coupling loss method, enhanced minimum coupling loss method, and the Monte-Carlo simulation. *European Radiocommunications*

*Committee Within the European Conference of Postal and Telecommunications Administrations*, (1999)

32. IEEE Std 802.16.2-2004, *IEEE Recommended Practice for Local and Metropolitan Area Networks Coexistence of Fixed Broadband Wireless Access Systems* (Institute of Electrical and Electronics Engineers, New York, 2004a)

33. V.K. Varma, H.W. Arnold, D. Devasirvatham, A. Ranade, L.G. Sutliff, Interference, sensitivity and capacity analysis for measurement-based wireless access spectrum sharing. IEEE Trans. Veh. Technol. **43**, 611–616 (1994)

34. NTIA Report 05-432, *Interference Protection Criteria Phase 1 - Compilation from Existing Sources,* (2005). Available: http://www.ntia.doc.gov/osmhome/reports /ntia05-432/ipc_ phase_1_report.pdf

35. FCC, *Spectrum Policy Task Force*. The interference protection working group. Frequency communication commission. (2002). From: http://www.fcc.gov/sptf/files/IPWGFinalReport. pdf

36. V.K. Varma, H.W. Arnold, D. Devasirvatham, A. Ranade, L.G. Sutliff, Interference sensitivity and capacity calculations for measurement-based wireless access spectrum sharing. Proc. IEEE Veh. Technol. Conf. 550–554 (1993)

37. IST-4-027756 WINNER II D 5.10.1, *The WINNER Role in the ITU Process Towards IMT-Advanced and Newly Identified Spectrum*, Vo1.0, November 2007

38. ITU-R WP 8F/975-E, *Analysis of Interference from IMT to FSS in the 3400–4200 MHz and 4500–4800 MHz*, Aug 2006

39. ITU-R Recommendation SF.1006, *Determination of the Interference Potential Between Earth Stations of the Fixed-Satellite Service and Stations in the Fixed Service*, Apr 1993

40. 3G Americas SM report, *3GPP Technology Approaches for Maximizing Fragmented Spectrum Allocations,* July 2009

41. R.K. Hamid, F. Martin, L. Gérard, F. Eric, European harmonized technical conditions and band plans for broadband wireless access in the 790–862 MHz digital dividend spectrum. *Conference of European Post & Telecommunications Administrations*. Singapore, 6–9 April 2010. pp. 978–987

42. CEPT Rep. 19, *Report from CEPT to the European Commission in Response to the Mandate to Develop Least Restrictive Technical Conditions for Frequency Bands Addressed in the Context of WAPECS*, October 2008

43. J. Laiho, A. Wacker, T. Novosad, *Radio Network. Planning and Optimisation for UMTS* (John Wiley & Sons, Sussex, 2002)

44. Z.A. Shamsan, A.M. Al-hetar, T.A. Rahman, Spectrum sharing studies of IMT-advanced and FWA services under different clutter loss and channel bandwidths effects. Prog. In Electromagn. Res., PIER **87**, 331–344 (2008)

45. WiMAX Forum (white paper), *Compatibility of Services Using WIMAX Technology with Satellite Services in the 2.3–2.7 GHz and 3.3–3.8 GHz Bands*, (2007)

46. RSAC Paper, *Assessment of Potential Interference Between Broadband Wireless Access Systems in the 3.4–3.6 GHz Band and Fixed Satellite Services in the 3.4–4.2 GHz Band*, Hong Kong, Feb 2006

47. Radiocommunication Study group, *Update of Sharing Studies Between IMT-Advanced and Fixed Satellite Service in the 3400–4200 and 4500–4800 MHz Bands*. Document 8F/927-E, August 2006

48. Recommendation ITU-R S.1432, *Apportionment of the Allowable Error Performance Degradations to Fixed-Satellite Service (FSS) Hypothetical Reference Digital Paths Arising from Time Invariant Interference for Systems Operating Below 15 GHz*, October 2008

49. APT Report, *The Co-Existence of Broadband Wireless Access Networks in the 3400–3800 MHz Band and Fixed Satellite Service Networks in the 3400–4200 MHz Band*. Draft no. APT/AWF/REP-5, March 2008

50. NTIA Report TR-99-361, *Technical Characteristics of Radiolocation Systems Operating in the 3.1–3.7 GHz Band and Procedures for Assessing EMC with Fixed Earth Stations Receivers*, December 1999

51. Radiocommunication Study Groups, *Experimental Evaluation on Robustness Against Potential Interference to TVRO Terminal from IMT-Advanced Transmitter in the 3400–4200 MHz Band*. Document 8F/1233-E, Japan, May 2007
52. IST-2006-027756 WINNER, *WINNER II Spectrum Sharing Studies*. D5.10.3 v1.0, (2007)
53. Radiocommunication Study group, *Sharing Studies Between FSS and IMT-Advanced Systems in the 3400–4200 and 4 500–4 800 MHz Bands*. R03-WP8F-C-1015, (2008)
54. L.F. Abdulrazak, A.O. Arshed, Interference mitigation technique through shielding and antenna discrimination. Int. J. Multimed. Ubiquit. Eng **10**(3), 343–352 (2015)
55. J.D. Gibson, *The Communications Handbook*, 2nd edn. (CRC Press, University of California, Santa Barbara, 2002)
56. Progira Rep, *Interference from Future Mobile Network Services in Frequency Band 790–862 MHz to Digital TV in Frequencies Below 790 MHz*. Progira Radio Communication AB Reprot, (2009)
57. Aegis (2007). *Digital Dividend (Interference Analysis of Mobile WiMAX, DTT and DVB-H Systems)*. Ofcom. 1913/DD/R2/3.0
58. J. Liberti Jr., T. Rappaport, *Smart Antennas for Wireless Communications: IS-95 and Third Generation CDMA Applications* (Prentice Hall PTR, Upper Saddle River, 1999)
59. J. Winter, Smart antennas for wireless systems. IEEE Pers. Commun. **5**(1), 23–27 (1998)
60. B. Widrow, P. Mantey, L. Griffiths, B. Goode, Adaptive antenna systems. IEEE Proc. **55**(12), 2143–2159 (1967)
61. H. Steyskal, Digital beamforming antennas, an introduction. Microw. J. **30**(1), 107–124 (1987)
62. Y. Karasawa, The software antenna: A new concept of kaleidoscopic antenna in multimedia radio and mobile computing era. IEICE Trans. Commun. **E80-B**(8), 1214–1217 (1997)
63. P. Lehne, M. Pettersen, An overview of smart antenna technology for mobile communications systems. IEEE Commun. Surv., Fourth Quarter **2**(4) (1999). http://www.comsoc.org/pubs/surveys
64. R. Ertel, P. Cardieri, K. Sowerby, T. Rappaport, J. Reed, Overview of spatial channel models for antenna array communication systems. IEEE Pers. Commun. Mag. **5**, 10–22 (1998)
65. B. Widrow, S.D. Stearns, *Adaptive Signal Processing* (Prentice Hall, Englewood Cliffs, 1985)
66. Z. Xinhua, C. Ling, C. Su-Shing, An experimental comparison between maximum entropy and minimum relative-entropy spectral analysis. IEEE Trans. Signal Process. **41**(4), 1730–1734 (1993)
67. B.D. Hart, Maximum likelihood sequence detection using a pilot tone. IEEE Trans. Veh. Technol. **49**(2), 550–560 (2000)
68. R. Schmidt, Multiple emitter location and signal parameter estimation. IEEE Trans. Antennas Propag. **34**(3), 276–280 (1986)
69. R. Roy, T. Kailath, ESPRIT - estimation of signal parameters via rotational invariance techniques. IEEE Trans. Acoust. Speech Signal Process. **ASSP-37**, 984–995 (1989)
70. J.S. Andreas, M. Lawrence Jr., S. Petre, Computationally efficient two-dimensional capon spectrum analysis. IEEE Trans. Signal Process **48**(9), 2651–2661 (2000)
71. C. Giorgio, L1-norm convergence properties of correlogram spectral estimates. IEEE Trans. Signal Process **55**(9), 4354–4365 (2007)
72. S. Petre, L. Jian, H. Hao, Spectral analysis of nonuniformly sampled data: a new approach versus the periodogram. IEEE Trans. Signal Process **57**(3), 843–858 (2009)
73. E.J. Paul, G.L. David, The probability density of spectral estimates based on modified periodogram averages. IEEE Trans. Signal Process **47**(5), 1255–1261 (1999)
74. A. Kuchar, M. Tangemann, E. Bonek, Real-time DOA-based smart antenna processor. IEEE Trans. Veh. Technol **51**(6), 1279–1293 (2002)
75. S. Anderson, M. Millnert, M. Viberg, B. Wahlberg, An adaptive array for mobile communication systems. IEEE Trans. Veh. Technol **40**(1), 230–236 (1991)

76. H.J. Thomas, T. Ohgane, M. Mizuno, A novel dual antenna measurement of the angular distribution of received waves in the mobile radio environment as a function of position and delay time. Proc. IEEE Veh. Technol. Conf, 546–549 (1992)

77. Y. Ogawa, N. Kikuma. High-resolution techniques in signal processing antennas. IEICE Trans. Commun.. **E78-B**(11), 1435 (1995)

78. S. Haykin (ed.), *Array Signal Processing* (Prentice-Hall, Englewood Cliffs, 1985)

79. J. Capon, High-resolution frequency-wavenumber spectrum analysis. Proc. IEEE **57**(8), 1408–1418 (1969)

80. S.M. Kay, *Modern Spectral Estimation: Theory and Application* (Prentice-Hall, Englewood Cliffs, 1988)

81. H. Krim, M. Viberg, Two decades of array signal processing research – The parametric approach. IEEE Signal Processing Mag **13**(4), 67–94 (1996)

82. D. Fuhrmann, L. Bede, Rotational search methods for adaptive Pisarenko harmonic retrieval. Acoustics Speech. Signal Proc. IEEE Trans **34**(6), 1550–1565 (1986)

83. W.F. Gabriel, Spectral analysis and adaptive array supper resolution techniques. Proc. IEEE **68**(6), 654–666

84. S.U. Pliiai, *Array Signal Processing* (Springer, New York Inc, 1989)

85. K. Takao, M. Fujita and T. Nishi. An adaptive antenna array under directional constraint. IEEE Trans. Antennas Propag. **AP-24**(5), 662–669 (1976)

86. R.T. Compton Jr., *Adaptive Antennas – Concepts and Performance* (Prentice-Hall, Englewood Cliffs, New Jersey, 1988)

87. R.T. Compton Jr., The power inversion adaptive array : concept and performance. IEEE Trans. Aerosp. Electron. Syst **AES-15**(6), 803–814 (1979)

88. R.L. Johnson, G.E. Miner, Comparison of super-resolution algorithms for radio direction finding. IEEE Trans. Aerosp. Electron. Syst **AES-22**(4), 432–442 (1986)

89. M. Kaveh, A.J. Barabell, The statistical performance of the MUSIC and the minimum-norm algorithms in resolving plane waves in noise. IEEE Trans. Acoust. Speech Signal Process. **ASSP-34**(2), 331–341 (1986)

90. P. Stoica, A. Nehorai, MUSIC, maximum likelihood, and Cramer-Rao bound. IEEE Trans. Acoust. Speech. Signal Process. **37**(5), 720–741 (1989)

91. B.D. Rao, K.V.S. Hari, Performance analysis of root-MUSIC. IEEE Trans. Acoust. Speech Signal Process. **ASSP-37**(12), 1939–1949 (1989)

92. M. Kim, K. Ichige, H. Arai, Implementation of FPGA based fast DOA estimator using unitary MUSIC algorithm. Proc. VTC 2003-Fall **1**, 213–217 (2003)

93. M. Kim, K. Ichige, H. Arai, Implementation of FPGA based fast unitary MUSIC DOA estimator. IEICE Trans **E87-C**(9), 1485–1494 (2004)

94. R. Roy, T. Kailath, ESPRIT—Estimation of signal parameter via rotational invariance techniques. IEEE Trans. Accoust Speech Proc. **37**, 984–995 (1989)

95. B. Ottersten, M. Viberg, T. Kailath, Performance analysis of the total least squares ESPRIT algorithm. IEEE Trans. on Signal Process. **SP-39**(5), 1122–1135 (1991)

96. M. Haardt, J.A. Nossek, Unitary ESPRIT : how to obtain increased estimation accuracy with a reduced computational burden. IEEE Trans. Signal Process. **43**(5), 1232–1242 (1995)

97. M.D. Zoltowski, M. Haardt, C.P. Mathews, Closed-form 2-D angle estimation with rectangular arrays in element space or beamspace via unitary ESPRIT. IEEE Trans. Signal Process. **44**(2), 316–328 (1996)

98. Y. Ogawa, N. Hamaguchi, K. Ohshima, K. Itoh, High-resolution analysis of indoor multipath propagation structure. IEICE Trans. Commun. **E78-B**(11), 1450–1457 (1995)

99. T. Kuroda, N. Kikuma, N. Inagaki, DOA estimation and pairing method in 2D-ESPRIT using triangular antenna array. IEICE Trans **J84-B**(8), 1505–1513 (2001)

100. Y. Tanabe, Y. Ogawa, T. Ohgane, High-resolution estimation of multipath propagation based on the 2D-MUSIC algorithm using time-domain signals. IEICE Trans **J83-B**(4), 407–415 (2000)

101. M. Zoltowski, G. Kautz, S. Silverstein, Beamspace root-MUSIC. IEEE Trans Signal Process. **41**(1), 344–364 (1993)
102. D. Kundu, Modified MUSIC algorithm for estimation DOA of signals, Elsevier. Signal Process **48**, 85–90 (1996)
103. A. Ferréol, P. Larzabal, M. Viberg, Statistical analysis of the MUSIC algorithm in the presence of modeling errors, taking into account the resolution probability. IEEE Trans. Signal Process. **58**(8), 4156–4166 (2010)
104. A. Ferréol, P. Larzabal, M. Viberg, On the resolution probability of MUSIC in presence of modeling errors. IEEE Trans. Signal Process. **56**(8), 1945–1953 (2008)
105. Y. Zhang, B.P. Ng, MUSIC-like DOA estimation without estimating the number of sources. IEEE Trans. Signal Process. **58**(3), 1668–1672 (2010)
106. T.B. Lavate, V.K. Kokate, A.M. Sapkal, Performance analysis of MUSIC and ESPRIT DOA estimation algorithms for adaptive array smart antenna in mobile communication. Int. J. Comput. Netw. (IJCN) **2**(3), 152–158 (2010)
107. P. Chargé, Y. Wang, J. Saillard, An extended cyclic MUSIC algorithm. IEEE Trans. Signal Process. **51**(7), 1695–1701 (2003)
108. Z. Shan, T.P. Yum, A conjugate augmented approach to direction-of-arrival estimation. IEEE Trans. Signal Process. **53**(11), 4104–4109 (2005)
109. X. Zhang, G. Feng, D. Xu, Blind direction of angle and time delay estimation algorithm for uniform linear array employing multi-invariance music. Progr. Electromagn. Res. Lett. **13**, 11–20 (2010)
110. Recommendations ITU-R BO.1213, *Reference Receiving Earth Station Antenna Patterns for Planning Purposes to be Used in the Revision of the WARC BS-77 Broadcasting-Satellite Service Plans for Regions 1 and 3*, October 2005
111. ETSI EN 301 390, *Fixed Radio Systems, Point-to-Point and Multipoint Systems, Spurious Emissions and Receiver Immunity Limits at Equipment/Antenna port of Digital Fixed Radio Systems*, December 2000
112. ECC Recommendation (04)05, *Guidelines for Accommodation and Assignment of Multipoint Fixed Wireless Systems in Frequency Bands 3.4–3.6 GHz and 3.6–3.8 GHz*, June 2005
113. WiMAX Forum, *WiMAX Deployment Considerations for Fixed Wireless Access in the 2.5 GHz and 3.5 GHz Licensed Bands*, June 2005
114. P. Kolodzy, Spectrum management with technical flexibility. *Invited Presentation to National Spectrum Managers Association Conference*, May 2008
115. Frank H. Sanders. *Measurements of Pulsed Co-Channel Interference in a 4-GHz Digital Earth Station Receiver*. NTIA Report 02–393, submitted to the U.S. DEPARTMENT OF COMMERCE, May 2002
116. Inter-American Telecommunication Commission report, *Technical Analysis of the Potential for Interference from Terrestrial Broadband Wireless Access ("BWA") Transmitters to Fixed-Satellite Service ("FSS") Receive Earth Stations in the Band 3,400–4,200 MHz*. During the meeting of permanent consultative committee, Lima, Peru, 20 June 2006
117. Radiocommunication Study group, *Sharing Study Between IMT-Advanced and Fixed Satellite Service in the Bands of 3400–4200 MHz and 4400–4990 MHz*. Document 8F/963-E, 15 Aug 2006
118. IST-2003-507581 WINNER, *Performance Assessment of the WINNER System Concept*. D7.8 V1.0, (2004)
119. IST-2003-507581 WINNER, *Duplex Arrangements for Future Broadband Radio Interface*. D2.5 V1.0, (2003)
120. IST-2003-507581 WINNER, *Winner System Concept Complexity Estimates*. D7.7 v. 1.0, (2004)
121. IST-2003-507581 WINNER, *Final Report on Identified RI Key Technologies, System Concept, and Their Assessment*. D2.10 v1.0, (2003)

# Chapter 3
# Interference Assessment Methodology

## 3.1  Introduction

The propagation model in the co-channel and adjacent channel interference scenario is calculated in this book based on traditional compatibility and sharing analysis method. The model consists of interferer adjacent channel leakage ratio (ACLR) and receiver adjacent channel selectivity (ACS). These two parameters will be incorporated into the wave propagation model in addition to the clutter loss effect. As a case study, one channel of MEASAT-3 C-Band downlink (36 MHz bandwidth) and frequency coordination with WiMAX service (20 MHz channel bandwidth) is considered for initial planning. The received power threshold is used as a benchmark for the interference between the two systems, while frequency offsets and geographical separations for different deployment environments are considered. This type of assessment has not been considered in any of the spectrum sharing studies reviewed in Chap. 2. The calculations are performed for 4 GHz frequency carrier, based on interference to noise ratio ($I/N$) of $-10$ dB.

MATLAB code (Appendix D) has been developed in order to obtain the specified value of $I/N$ by tuning up the minimum separation distance to an appropriate level that corresponds to a set of frequency offsets between carriers. Finally, Transfinite Visualyse Pro™ is used to verify the results. (Details about Visualyse Pro™ are attached in Appendix E).

## 3.2  Assessment Methodology

The research assessment methodology flowchart in Fig. 3.1 represents the main procedures required to execute the research objectives.

© Springer International Publishing AG 2018
L.F. Abdulrazak, *Coexistence of IMT-Advanced Systems for Spectrum Sharing with FSS Receivers in C-Band and Extended C-Band*,
https://doi.org/10.1007/978-3-319-70588-0_3

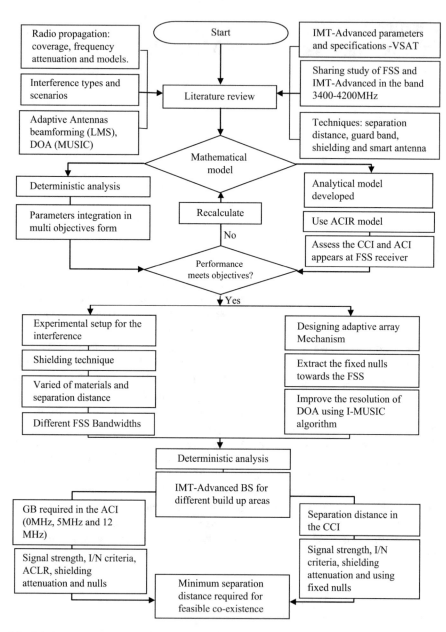

**Fig. 3.1** Research methodology

As shown in Fig. 3.1, in this book, the literature reviews have developed on different related matters such as band segmentations, sharing, frequency allocation, interferences, ACLR model, shielding and smart antenna mitigation technique. Afterwards, IMT-Advanced (proposed system WiMAX IEEE802.11e) and FSS receiver specifications have been determined to identify the protection ratio, which is depended on the system bandwidths. Then, separation distance in CCI and ACI scenarios will be easily calculated using the derived propagation model. This propagation model will develop a different assumption to introduce the ACIR within the guard band analysis.

Evaluate the received signal by FSS via MEASAT-3, in order to check on the effect of different shielding materials. Therefore, the best shielding material and scenario will be concluded in Chap. 4 after the experiment setup. Subsequently, determining the intersystem interference scenarios in different deployment areas according to the requirements will be carried out in the same chapter.

The problem of accurate direction of arrival detection using Improved MUSIC (I-MUSIC) algorithm will be presented in Chap. 5. Then, the results will be verified by simulation. Comparisons between the root mean squared error (RMSE) and computational complexity costs against those of conventional algorithms are performed. Numerical results for different array antenna manifolds and a variety of data lengths are also presented in the simulations.

Last step in the methodology would be proposing a complete mechanism for beam cancellation system, to coexist the IMT-Advanced base stations (BS) and FSS in the 3400–4200 MHz frequency range. This mechanism should be able to steer the radiation power towards the user while still fixing the fixed nulls in the direction of FSS receiver. Finally, coexistence scenarios will be determined, and the feasibility of the proposed mitigation techniques on the coordination process will be evaluated.

## 3.3  Propagation Model Parameters

Literature review has shown the importance of standard model agreed upon in CEPT and ITU for a terrestrial interference assessment at microwave frequencies, which includes the attenuation due to clutter in different environments. In reviewing the literature, no data has been found on the association between clutter loss and frequency separation which represents the core of interference avoidance study. Therefore, this study is aimed at assessing the importance of radio propagation coverage and models.

The MCL method is a straightforward approach as it allows the radio engineer to quickly determine the minimum frequency separation by using some worst-case assumptions. The disadvantage of this simplicity is a reduction in spectrum efficiency. By using worst-case assumptions, the minimum frequency separation is significantly greater than the required in the practice [1]. The ITU-R452.14 propagation model is used in the terrestrial communications to account for the deteriorating effects of clutters on radio transmissions at frequencies above 0.7 GHz in different deployment environments. Remarkably, it has been found that WiMAX 802.16e operating environments (dense urban, urban, suburban and rural) have several similar characteristics such as interference to noise ratio.

Intuitively by introducing clutters, smaller separation distance is achieved and vice versa. However, due to the high sensitivity of FSS receiver, transmission footprint is much larger than normal receivers operating in the same area. Path loss prediction in the case of line of sight (LOS) is obtained by including the losses produced by the line-of-sight situation together with the losses produced by clutter models. This is summarized in Eq. (3.1) [2]:

$$L(d) = 92.44 + 20\log_{10} f_{GHz} + 20\log_{10} d_{Km}$$
$$+ 10.25 e^{-d_k}\left[1 - \tanh\left[6\left(\frac{h}{h_a} - 0.625\right)\right]\right] - 0.33 \tag{3.1}$$

where $d$ is the distance between the interferer and the victim receiver in kilometres, $f$ is the carrier frequency in Gega hertz and $A_h$ is loss due to protection from local clutter or called clutter loss and $d_k$ is the distance in km from nominal clutter point to the antenna ($d_k = 0.02$ km, 0.02 km, 0.025 km and 0.1 km for the four deployment environments dense urban, urban, suburban and rural, respectively), $h$ is the antenna height (m) above local ground level and $h_a$ is the nominal clutter height above local ground level ($h_a = 25$ m, 20 m, 9 m and 5 m for the four deployment environments). The separation distance can be calculated as follows:

$$20\log(d) = -I + EIRP_{Interferer} - 92.5 - 20\log(F) - A_h + G_{vs}(\alpha) \tag{3.2}$$

where EIRP is the effective isotropic radiated power transmitted from the interferer, $F$ is the frequency in GHz and $G_{vs}$ is related to the typical receiving FSS antenna gain [3]. Equation (3.2) accounts for the most important parameters affecting the radio propagation, which might also apparently be subdivided into other subparts.

For realistic consideration of interference, the ACLR and ACS are included and derived from Eq. (3.2). The receiving gain of FSS station is called off-axis antenna $G_{vs}(\alpha)$. The off-axis angle value depends on the earth station location and the main receiving beam, where a typical receiving antenna gain can be calculated as Eq. (3.3) [4]:

$$G_{vs}(\alpha) = \begin{cases} G_{max} - 2.5 \times 10^{-3}\left(\dfrac{D}{\lambda}\phi\right)^2 & 0 < \alpha < \phi_m \\ 52 - 10\log_{10}\left(\dfrac{D}{\lambda}\right) - 25\log_{10}(\alpha) & 3.6° < \alpha < 48° \\ -10\,dBi & 48° < \alpha < 180° \end{cases} \tag{3.3}$$

where $G_{max}$ is the maximum antenna gain (38dBi), $D = 1.8$ m (satellite diameter) and $\lambda$ is the wave length in metre and $\phi_m$ is given by

$$\phi_m = \frac{20\lambda}{D}\sqrt{G_{max} - 2 - 15\log_{10}\left(\frac{D}{\lambda}\right)} \qquad (3.4)$$

In the simulation a value of −10 dB was considered to represent the local case study (the FSS elevation angle at the experiment location was 74°).

## 3.4   Receiver Blocking and Adjacent Channel Interference Ratio (ACIR) Coexistence Model

In addition to the deterministic approach presented in the previous section, other critical parameters, such as the receiver blocking and ACIR, are considered in this work. These resulted from the introduction of the spectrum emission mask (SEM) of the interferer and the blocking filter capability of the victim. Thus, this is reflected in the actual nature of the interference between the two types of base stations. It is worth mentioning that receiver blocking and ACIR calculations are based on transmitter SEM and victim filter response powers [5].

The SEM and victim filter response powers are divided into a number of segments with power spectral density (PSD) attenuation on y-axis and frequency spacing on x-axis. The PSD is governed by the regulators in order to reduce the harmful interference. In line with the above context, which is typical of the co-channel interference (CCI) scenario, the receiver blocking is considered in order to find the power degradation in decibel. This can be calculated as follows:

$$\text{Receiver Blocking} = \begin{cases} \text{ACS Reduction} & \text{if} \quad BW_{interferer} < BW_{Victim} \\ 0\,dB & \text{if} \quad BW_{interferer} \geq BW_{Victim} \end{cases} \qquad (3.5)$$

where ACS reduction is the value of received signal within the interference, for the case when the interferer bandwidth is less than the victim bandwidth. It is worth mentioning here that if both signals have the same bandwidth, ACS calculation will not be necessary and only the SEM of the interferer will rise to interference power in the case of CCI. Such a SEM is the 20 MHz channel bandwidth type-G WiMAX spectrum emission mask in [1].

WiMAX system featuring this type of SEM was envisaged as the next-generation technology and is chosen for coexistence and sharing studies. Table 3.1 shows some typical values for mobile WiMAX's SEM.

**Table 3.1** Reference frequency for SEM of type-G (WiMAX) [1]

| Channel separation (MHz) | 0 | 0.5 | 0.5 | 0.71 | 1.06 | 2 | 2.5 |
|---|---|---|---|---|---|---|---|
| WiMAX band width from $Fc$ (MHz) | 0 | 10 | 10 | 14.2 | 21.2 | 40 | 50 |
| Power loss (dB) | 0 | 0 | −8 | −32 | −38 | −50 | −50 |

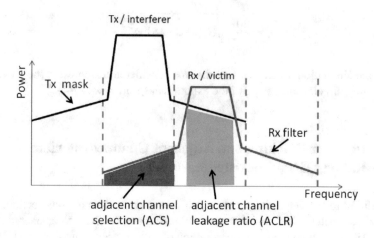

**Fig. 3.2** ACLR and ACS as issues in the transmitter mask and receiver filter [6]

The channel 20 MHz bandwidth is multiplied by a factor 0.5, which is the normalized frequency offset, in order to achieve a 10 MHz separation from the assigned frequency carrier. The power spectrum attenuation should be 8 dB [1]. At the same time, all the frequency offsets and the corresponding power spectral densities will conform to the following straight-line equation:

$$\Delta f = a\left(\Delta f\right) + b \tag{3.6}$$

where $\Delta f$ denotes the frequency offset from the carrier, $a$ represents the amount of attenuation in dB in the segment and $b$ is the attenuation in dB at a certain frequency offset of $f$ from the reference. The SEM will determine the values of ACLR depending on the frequency offsets between the interferer and victim channels. As the frequency offset becomes wider, the effect of ACLR will be less. Therefore, it is important to consider the power loss due to the channel separation in the SEM. Figure 3.2 shows the unwanted emission mask intersecting with the receiver filter causing an ACI scenario.

In Fig. 3.2 the vertical line intersecting with the grey area indicates the leakage in the transmitter power value which can affect the victim received power. Therefore, as the victim carrier moves away from the interferer carrier, the SEM power will be reduced, and corresponding sufficient guard band can be calculated.

The FSS channel selectivity is obtained by superimposing the front-end bandpass filter on an assumed typical IF (70 MHz) surface acoustic wave filter (36 MHz bandwidth). So, ACS is the ratio of receiver filter attenuation over its pass-band to by the receiver's filter attenuation over an adjacent frequency channel. As an example for practical measurements of the ACS values, Table 3.2 represents the channel bandwidth and the amount of power reductions as the signal deviated from the frequency carrier [7].

**Table 3.2** FSS receiver channel selectivity attenuation front-end plus IF

| Channel separation (MHz) | 0 | 0.5 | 0.58 | 0.66 | 1.38 | 1.66 | 2.5 |
|---|---|---|---|---|---|---|---|
| WiMAX bandwidth from $f_c$ (MHz) | 0 | 18 | 21.1 | 24 | 50 | 60 | 90 |
| Power loss (dB) | 0 | 0 | −40 | −48 | −61 | −66 | −75 |

Table 3.2 shows that the power reduction corresponds to the amount of separation distance from the frequency carrier, which correspondingly increases the selectivity of adjacent channel. The ACS reduction at the CCI scenario and the victim bandwidth can be relatively calculated in different ways. Using the power attenuation in Tables 3.1 and 3.2, the ACS reduction reaches up to 34.5 dB for the FSS (36 MHz) and WiMAX (20 MHz) channel bandwidths. This value can be used when the frequency offset is 0 MHz.

In order to calculate the adjacent channel interference, the ACIR should be considered by reducing the interference powers of the interferer ACLR and the victim ACS which are located on different central frequencies [6]. The ACIR represents the relationship between the total power received from the interferer leakage and the receiver channel selectivity at specific channel separation. The ACIR is given by

$$ACIR = \frac{1}{\dfrac{1}{ACLR} + \dfrac{1}{ACS}} \tag{3.7}$$

where ACIR is the ratio of the interference power on the adjacent channel to the interference power experienced by the victim, ACLR is the ratio of the power over signals pass-band to the interference power over receiver pass-band and ACS is the ratio of the receiver pass-band attenuation to the receiver filter adjacent channel attenuation.

Therefore, the interference power can be calculated as a summation of out-of-band interference and in-band interference. In order to calculate the separation distance required to achieve the coexistence between two systems in ACI scenario, it is important to include the ACIR as follows:

$$20\log_{10}(d) = -I + P_t + G_{tAnt} + G_{tBF} + G_{vs}(\alpha) - 92.44 - 20\log_{10}(F) - A_h - R$$
$$+ ACIR + Corr\_B \tag{3.8}$$

where the Corr_B is the band correction that is equivalent to zero if the interferer bandwidth is less than victim bandwidth; and it is equal to 10 Log (interferer bandwidth/victim bandwidth) if the victim bandwidth is less or equal to the interferer bandwidth [7]. The ACIR values are calculated as follows:

$$ACIR = 10\log_{10}\left[\left(10^{ACLR_{IMT}/10}\right)^{-1} + \left(10^{ACS_{FSS}/10}\right)^{-1}\right]^{-1} \tag{3.9}$$

**Table 3.3** ACIR values at the adjacent channel interference scenario

| ACI scenario | $\Delta F$ (frequency offset) | ACS (dB) | ACLR (dB) | ACIR (dB) |
|---|---|---|---|---|
| 1st adjacent channel | 28 MHz | −50 | −42 | −50.638 |
| 2nd adjacent channel | 33 MHz | −53 | −45 | −53.639 |
| 3rd adjacent channel | 40 MHz | −56 | −50 | −56.97 |

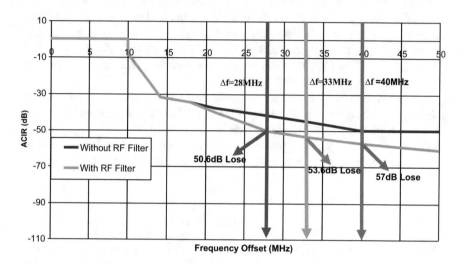

**Fig. 3.3** Compression between ACLR and ACIR

Using Tables 3.1 and 3.2, ACLR and ACS can be obtained for any frequency offset within the proposed range. The ACIR can be calculated using Eq. (3.9) which depends on the receiver ACS and transmitter ACLR. The results are specified in Table 3.3.

Table 3.3 shows the power loss due to the ACIR between the WiMAX transmitted signal and the FSS receiver filter. In order to calculate the interference in the adjacent channel, the ACIR should be considered by reducing the interference powers of the interferer and the victim that are located in different central frequencies [6]. As a result the ACIR has been calculated and obtained in Fig. 3.3 for $\Delta f = 28$, 33 and 40 MHz based on ACS and ACLR.

Figure 3.3 shows that frequency offsets have been chosen to get guard bands of 0, 5 and 12 MHz, respectively. Clearly, the ACS of FSS receiver plays an important role in reducing the harmful adjacent channel leakage of the receiver by 8.6, 8.6 and 7 dB, respectively, when $\Delta f = 28$, 33 and 40 MHz. Using ACIR offers other good features on the analytical model. These features include the same method for studying the interference in the larger guard band and the ability to obtain the coexistence for a required separation distance by controlling the frequency offset between two systems. Other features are the ability to obtain the coexistence for a frequency offset between the two systems by controlling the separation distance and the ability to investigate coexistence between two systems having the same or different channel bandwidths for both interferer and victim.

## 3.5   Interference Assessment Scenarios

The three scenarios of interference are depicted in Fig. 3.4, where the first scenario shows the interfering signal falling within the operating band of the victim FSS receiver (i.e. they are co-channelled). In the second scenario, both victim and interferer bands fall outside each other and are located contiguously (adjacent). Nevertheless, inserting a guard band in between the two systems forms the third coexistence scenario, where interference is therefore minimal.

In each of the interference scenarios, there is the need to consider a certain procedure in order to achieve the coexistence. However, unwanted emissions such as out of band (OOB), which is 250% of the spurious emission of the interfering equipment, fall within the receive band of the victim receiver and thereby acting as co-channel interference to the wanted signal [1]. The victim receiver blocking is defined as the capability of receiver to block a strong signal out of the receiver band in order to protect the reception desensitization. In general, this sort of interference can only be removed at the victim. However, in most cases the adoption of power control for the interferer and good site engineering can improve the situation. In order to compare these scenarios, the same propagation model should be adopted for all methods.

For the purpose of analysing the coexistence in various environments, these three scenarios will be used and simulated based on the aforementioned deployment areas. By considering the channelization plan for MEASAT downlink transponders, one channel is proposed at 4 GHz central frequency as illustrated in Fig. 3.5.

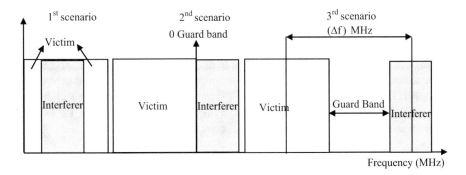

**Fig. 3.4**   The coexistence of co-channel, zero-guard band and adjacent channel

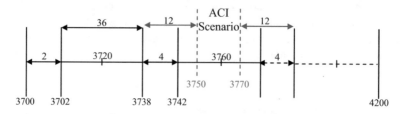

**Fig. 3.5** MEASAT-3 Spectrum plan with the proposed channel

Figure 3.5 depicts the downlink channelization plan for MEASAT-3 geostationary satellite orbit where the proposed 20 MHz WiMAX replaces the 36 MHz FSS channel. Consequently, a maximum of 12 MHz guard band may be obtained when such a proposal is applied. This outcome was actually the driving impetus behind the specified frequency offset. A fixed protection ratio has been used as a criterion for the interference scenarios in order to limit the interference, by using $I/N = -10$ dB. Note that $I$ is the interference level in dBm, $N$ is the noise floor of receiver in dBm and $-10$ dB is the protection ratio for FSS receiver.

Actually the 10 dB has been obtained by using degradation factor of 0.5 for carrier to noise ratio. Therefore, the value of interference level below noise floor is

$$\text{Interference below noise floor} = 10\log_{10}\left(1/\left(10^{0.5/10} - 1\right)\right)\text{dB} = 10\text{dB} \quad (3.10)$$

MEASAT satellite has 24 transponders, and each of the transponders has a maximum of 36 MHz channel bandwidth. In order to determine the maximum possible level of in-band interference at the FSS receiver, the following expressions have been used:

$$\frac{C}{I} = \frac{I}{N} + \frac{C}{N} = (10 + 5.7)\text{dB} \quad (3.11)$$

$$C = \frac{C}{N} + 10\log_{10}(KTB)\text{dBw} \quad (3.12)$$

$$I_{\text{inband}} = (C - 15.7)\text{dBw} = -143\text{dBw} \quad (3.13)$$

where $C$ is the carrier power at the receiver in dB, $C/N$ is the required carrier to noise ratio which is specified as a 5.7 dB minimum [4], $I_{\text{in-band}}$ is the required protection ratio, $K$ is Boltzmann constant = $1.38 \times 10^{-23}$ J/k, $T$ is the temperature in Kelvin and $B$ is the noise bandwidth in Hz [8]. With the carrier frequency at 4 GHz, the overall propagation model may be rewritten as follows:

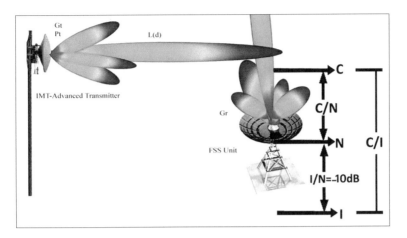

**Fig. 3.6** The simulated scenario between WiMAX BS and FSS ES

$$20\log_{10}(d) = -I + \text{EIRP}_{\text{WiMAX}} + G_{vs}(\alpha) - 104.58 + \text{ACIR} + \text{corr\_band}$$
$$- \left(10.25e^{-d_k}\left[1 - \tanh\left[6\left(\frac{h}{h_a} - 0.625\right)\right]\right] - 0.33\right) \quad (3.14)$$

The ACIR is the adjacent channel interference ratio due to the receiver filter and transmitted mask at any frequency offset. Corr_band is the correction factor of the band ratio, which is equal to 0 dB when $\text{BW}_{\text{WiMAX}} < \text{BW}_{\text{FSS}}$. Otherwise, Corr_band $= -10\log(\text{BW}_{\text{WiMAX}}/\text{BW}_{\text{FSS}})$, when $\text{BW}_{\text{WiMAX}} > \text{BW}_{\text{FSS}}$. Therefore, when the bandwidth of FSS is 230 kHz, the correction band is given by the following expression:

$$\text{Corr\_band} = -10l\log_{10}\left(\frac{20 \times 10^6 \,_{\text{WiMAX}}}{230 \times 10^3 \,_{\text{FSS}}}\right) = -19.4\text{dB} \quad (3.15)$$

The worst case for the sharing scenario between WiMAX 802.16e and FSS is simulated. Figure 3.6 shows the different parameters used in the simulation process, when the IMT-Advanced antenna is facing the FSS ES.

Analysis of the received interference at the victim FSS receiver involves dividing the interference into noise level and interference criterion, as indicated in the following expression:

$$\text{INR} = \text{EIRP}_{\text{WiMAX}} + G_r - 104.58 + \text{ACIR} + \text{corr\_band} - 20\log(d) - N$$
$$- \left(10.25e^{-d_k}\left[1 - \tanh\left[6\left(\frac{h}{h_a} - 0.625\right)\right]\right] - 0.33\right) \quad (3.16)$$

where INR is the FSS victim receiver's interference to noise ratio. Different environments are considered in the simulation process, and the results are briefly discussed in the following sections.

## 3.6  Minimum Separation Distance Simulation

This study is aiming to define the minimum separation distance between WiMAX base station (the interferer) and FSS receiver (the victim). To achieve this objective, the parameters of proposed system are specified, and the possible factors that may affect the frequency coordination are evaluated. Such factors include victim gain, interferer ACLR, receiver ACS, antenna heights and bandwidths.

The proposed area analysis gives an insight into the massive interference that may impact the performance of FSS receiver in different environments. Two types of interference will be discussed for the deployment areas.

**Table 3.4**  WiMAX 802.16 systems parameters

| Parameter | Value WiMAX |
|---|---|
| Centre frequency of operation (MHz) | 4000 |
| Channel bandwidth (MHz) | 20 |
| Base station transmitted power (dBm) | 43 |
| Spectral emissions mask requirements | ETSI-EN301021Type G |
| Base station antenna gain (dBi) | 18 |
| Base station antenna height (m) | 30 |

**Table 3.5**  Fixed-satellite service specifications

| Specifications | Satellite terminal |
|---|---|
| Antenna diameter (m) | 1.8 |
| Antenna gain (dBi) | 38 |
| Frequency $F_c$ (MHz) | 4000 |
| Elevation angle | $74°$ |
| Azimuth | $263.7°$ |
| Height (m) | 1.8 and 5 |
| Received noise temperature | 114 K |
| Bandwidth (MHz) | 0.23 and 36 |
| Theoretical interference level (I) | −165 dBw/0.23 MHz<br>−143 dBw/36 MHz |

### 3.6.1 Simulated Parameters

A simulation of two base stations, WiMAX 802.16e and FSS, has been conducted by proposing the most appropriate parameters for WiMAX 802.16e and the used parameters for the FSS receiver. These parameters are quite compatible with those proposed in the WRC-07 [9]. Tables 3.4 and 3.5 specify the WiMAX and FSS parameters, respectively.

For WiMAX base station, it is proposed to be macro station coverage with a SEM type-G (model number: UTIS EN301021), with the antenna height fixed to 30 m above ground level. Antenna gain proposed to be 18dBi for three sector base stations, and power transmitted of 43 dBm is applied for the WiMAX systems. The frequency carrier is attuned in order to benefit the coexistence in the adjacent channel interference. A really rare case is considered when co-channel interference happens at the 4000 MHz frequency carrier. It is important to note that these parameters are based on proposed scenario parameters.

The FSS parameters in Table 3.5 reflect the fact that a fixed position of FSS receiver is used, while the WiMAX ones apparently show movable type of WiMAX base station parameters. Parameters in Tables 3.4 and 3.5 are used to assess the interference between WiMAX and FSS in terms of minimum separation distances using the proposed propagation model of Eq. (3.16). A FSS antenna of variable heights (1.8 m and 5 m) is used to highlight the fact that the position of the FSS receiver on the ground can reduce the separation. For the FSS receiver, it is recommended to keep the dish on the ground level to improve the coexistence. Although based on the literature, it has been noticed that some users locate their FSS antennas on the roof of first floor. Referring to the effect of clutter loss, it is noticed that increasing the FSS receiver height corresponds to extending the separation distance. Therefore, if the FSS base station is higher than the clutter height, then the minimum distance required will remain fixed.

### 3.6.2 Protection Ratio Methodology

The calculation for protection ratio ($I/N$) can be summarized as shown in Fig. 3.7. According to the figure, in order to make proper frequency coordination, the resultant protection ratio should be compared with $I/N$ obtained in the input of victim receiver.

Figure 3.8 presents the flowchart of three phases whereby the frequency coordination is achieved between services by adjusting the frequency (or geographical site coordination), based on $I/N$ protection ratio. The first phase demands for the system parameters which serve as the inputs to the proposed propagation formulas. These formulas are the program engines which identify whether or not the parameters are working correctly.

The second phase divides the interference scenarios into two parts (either co-channel interference or adjacent channel interference) when the frequency offset

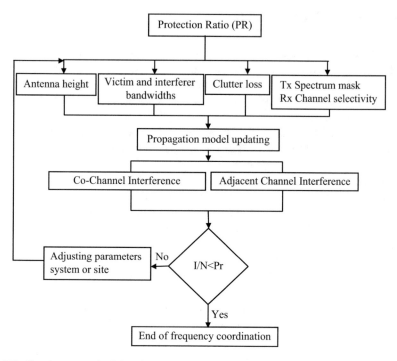

**Fig. 3.7** Coexistence methodology in response to the protection ratio

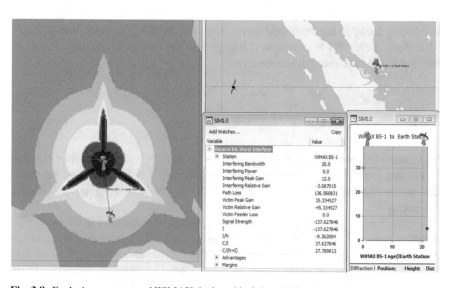

**Fig. 3.8** Exclusion zone around WiMAX deployed in Johor, Malaysia

exists between frequency carriers. Eventually, the desired received signal strength
(DRSS) and interference received signal strength (IRSS) level are compared to verify whether protection ratio is secured or not [6].

When the interference occurs, a protection ratio can also be identified using the
carrier to interference ratio (*C/I*) or carrier to noise plus interference ratio (*C/N + I*).
In both cases the DRSS or IRSS should be above the protection ratio to avoid the
interference. These protections can be expressed as follows:

$$\frac{C}{N+I} = \frac{C}{I} - \frac{N+I}{N} - \frac{I}{N} \tag{3.17}$$

where $C/(N + I)$ is the value on carrier above the noise level with present of interference, $(N + I)/N$ is the desensitization level and $I/N$ is the new proposed protection
ratio.


### 3.6.3   Area Analysis

Visualyse Professional can be used to analyse a very wide range of scenarios. These
include studies which cover different types of stations and services, including
WiMAX, mobile, fixed, GSO satellite and FSS receiver, to mention but a few.

In this section a discussion on WiMAX BS deployment in the co-frequency area
of an existing fixed-satellite service (FSS) earth station is discussed. However, area
analysis can be used to show these colour-coded plots or contours of any link parameters (e.g. received noise, interference, $I/N$, PFD). So, the next phase is to repeat the
analysis undertaken at a single point (above) over an area and showing its variation.
This is done by using the Visualyse Professional Area Analysis tool.

There are two particular ways in which the area analysis can be used, as
concluded:

1. Varying the location of the WiMAX base station to investigate the variation of
   interference at the FSS earth station with location and defining exclusion zones
   around the earth station where WiMAX could not be deployed
2. Varying the location of the FSS earth station to investigate the variation of
   received interference with location and hence defining the exclusion zones where
   an FSS earth station could not be deployed because of a WiMAX base station

In this case the second option was chosen to vary the location of the earth
station.

An area analysis is selected from the map view with the following properties: the
moving station is proposed to be the FSS earth station, which represented the victim
link, attribute to plot is the aggregate interference $I/N$, plot resolution is 1 km and
the colour display settings were $I/N \leq -20$ dB is not shown, $-20$ dB $\leq I/N \leq -10$ dB
is shown as green, $-9$ dB $\leq I/N \leq 0$ dB is shown as orange, 1 dB $\leq I/N \leq 5$ dB is
shown as yellow, 5 dB $\leq I/N \leq 10$ dB is shown as red and $I/N \geq 10$ dB is shown as
brown.

Figure 3.8 shows the exclusion zone that was obtained from basic analyses under the assumption of path loss occurring for 20% of the time in Recommendation ITU-R 452-12. When using other percentages of time, it should be remembered that the $I/N$ threshold will also change spot which could increase or decrease the interfered area.

Figure 3.8 shows the thermal image of area analysis, where the protection ratio is represented by the green colour, where $I/N = -10$ dB. A complete scenario of both systems is represented on the right side of Fig. 3.8. Accordingly the separation distance was 23 km for the proposed parameters as highlighted in the window. It can be seen that a large area is excluded.

Note that the FSS pattern is not symmetric because it points to the satellite at longitude of $103°E$ which is towards the South East. It has slightly lower gain towards the WiMAX base station.

### 3.6.4   Separation Distance Between WiMAX and FSS of 36 MHz

Mathematical expression for the deterministic model explained in Eq. (3.16) was also used to determine the relationship between the separation distance and the required guard band. An important role played by the interference to noise ratio, during the simulation, is to represent the ACIR for different scenarios within all the deployment areas.

The results of separation distance have been summarized in Fig. 3.9, when the values of frequency offset between carriers are 0, 28 and 40 MHz. By fixing the limit of $I/N$ to $-10$ dB as a protection ratio, the calculations for co-channel and adjacent channel are performed for the current frequency band 4GHz (these include interferers' adjacent channel leakage ratio (ACLR) and receivers' adjacent channel selectivity (ACS)).

Figure 3.9 shows the separation distance between FSS ES and WiMAX in four different deployment areas, when a FSS antenna height of 5 m is used for CCI and ACI scenarios using MATLAB simulation results.

In order to verify the results obtained, the same scenarios are rebuilt into Transfinite Visualyse Pro™. This software is based on a number of statistical and deterministic approaches developed for best coexistence practices between two similar or dissimilar systems.

As also shown in Fig. 3.9, the simulated results of the Visualyse software are almost compatible with MATLAB simulation. This proves that the ACIR model is useful for high accuracy in terms of coordination between systems. The results obtained also show a small variance according to the ducting signal factor considered by Visualyse Pro™.

The impact of different deployment areas was highlighted in Fig. 3.9. It shows that nature of clutter could affect the overall results. It can be used for different

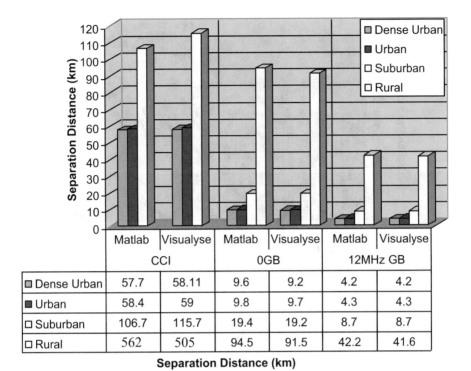

| | Matlab | Visualyse | Matlab | Visualyse | Matlab | Visualyse |
|---|---|---|---|---|---|---|
| | CCI | | 0GB | | 12MHz GB | |
| ▨ Dense Urban | 57.7 | 58.11 | 9.6 | 9.2 | 4.2 | 4.2 |
| ▣ Urban | 58.4 | 59 | 9.8 | 9.7 | 4.3 | 4.3 |
| □ Suburban | 106.7 | 115.7 | 19.4 | 19.2 | 8.7 | 8.7 |
| □ Rural | 562 | 505 | 94.5 | 91.5 | 42.2 | 41.6 |

**Separation Distance (km)**

**Fig. 3.9**  Results of separation distance using the MATLAB and Visualyse simulation for four different environments at different frequency offsets

deployment environments which justified the effects of using different antennas, clutter heights and nominal building separations.

A minimum separation distance between two services was achieved when the dense urban environment was selected for the deployment area. The worst case of interference occurs when the rural deployment area is used. A minimum separation distance without mitigation technique was endorsed when 12 MHz is used as a guard band in urban and dense urban area.

Therefore, the minimum separation distance was credited for variable FSS antenna heights using MATLAB program. In order to determine the effect of the antenna heights, different deployment areas are considered with respect to the CCI and adjacent channel interference scenarios as shown in Fig. 3.10. However, the frequency offsets were equal to 0, 28, 33 and 40 MHz for 36 MHz FSS channel bandwidth.

However, the clutter loss is directly related to the antenna heights. Therefore, for an antenna height of 1.8 m, clutter height losses are 19.74 dB, 19.73 dB, 19.5 dB and 15.4 dB as isolation provided in the dense urban, urban, suburban and rural areas, respectively. On the other hand, when the FSS height is 5 m, then the corresponding clutter isolation will be 19.65 dB, 19.55 dB, 13.6 dB and −0.32 dB, respectively.

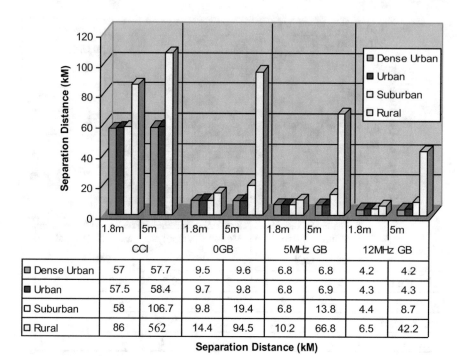

| | CCI | | 0GB | | 5MHz GB | | 12MHz GB | |
|---|---|---|---|---|---|---|---|---|
| | 1.8m | 5m | 1.8m | 5m | 1.8m | 5m | 1.8m | 5m |
| Dense Urban | 57 | 57.7 | 9.5 | 9.6 | 6.8 | 6.8 | 4.2 | 4.2 |
| Urban | 57.5 | 58.4 | 9.7 | 9.8 | 6.8 | 6.9 | 4.3 | 4.3 |
| Suburban | 58 | 106.7 | 9.8 | 19.4 | 6.8 | 13.8 | 4.4 | 8.7 |
| Rural | 86 | 562 | 14.4 | 94.5 | 10.2 | 66.8 | 6.5 | 42.2 |

**Separation Distance (kM)**

**Fig. 3.10** Separation distance using variable FSS antenna heights for four different environments at different frequency offsets

By referring to the propagation model, the clutter loss in the rural area coverage must be set to a value less than zero. These values will definitely increase the separation distances, and consequently it is most difficult to achieve the coexistence in the rural area, compared to other deployment areas.

According to Fig. 3.10, these values indicate that receiver victim heights play important roles in the clutter loss calculations to determine the amount of interference. Figures 3.9 and 3.10 summarize the results of separation distance which are based on MATLAB code attached in Appendix D.

### 3.6.5 Separation Distance Between WiMAX and FSS at 0.23 MHz

When a different FSS receiver is used in the simulation, the interference level will change in response to the victim bandwidth as earlier mentioned in Eq. 3.10. Accordingly, when bandwidth of 0.23 MHz is used to conduct the assessment, the interference level was −165 dBW/0.23 MHz as indicated in Table 3.5. It can produce a longer separation distance due to the increase in the receiver sensitivity. In the analytical study considered where the FSS antenna heights are 1.8 m and 5 m, different values of separation were obtained due to the clutter loss.

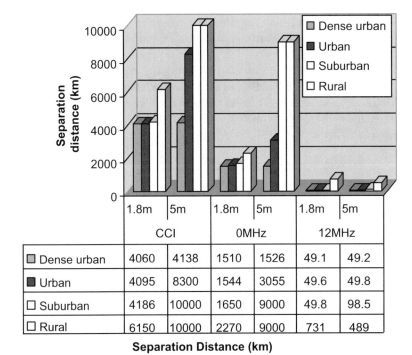

| | 1.8m | 5m | 1.8m | 5m | 1.8m | 5m |
|---|---|---|---|---|---|---|
| | CCI | | 0MHz | | 12MHz | |
| ▣ Dense urban | 4060 | 4138 | 1510 | 1526 | 49.1 | 49.2 |
| ■ Urban | 4095 | 8300 | 1544 | 3055 | 49.6 | 49.8 |
| ☐ Suburban | 4186 | 10000 | 1650 | 9000 | 49.8 | 98.5 |
| ☐ Rural | 6150 | 10000 | 2270 | 9000 | 731 | 489 |

**Separation Distance (km)**

**Fig. 3.11** The separation distance between WiMAX and FSS when FSS bandwidth is 0.23 MHz in the four deployment areas for different frequency offsets

When the victim bandwidth is less than the interferer bandwidth, the sufficient separation distance for a specific guard band can be tuned manually. Even though without incorporating a mitigation technique in this scenario, the guard band required to protect the victim is efficient. Therefore, it is more practical to determine the required separation distance directly after obtaining the power reduction from the ACLR.

The interference from WiMAX (20 MHz bandwidth) to FSS (0.23 MHz bandwidth) was simulated, and interference to noise protection ratio of −10 dB was considered. Thus, the scenarios of CCI and ACI are considered to represent the frequency dimension as a possible mitigation technique. This is used to reduce the separation distance between the two services as presented in Fig. 3.11.

Figure 3.11 shows that minimum required separation distance of 49.1 km could be achieved with 1.8 m FSS receiver height in the ACI scenario under dense urban area environment. However, the worst case of interference is experienced when a rural area was used as the deployment scenario in the CCI. The minimum separation distances required for the urban area deployment were 4.3 km and 49.6 km for 36 MHz and 0.23 MHz FSS channel bandwidths, respectively. Note that a 12 MHz guard band and 1.8 m antenna height were used.

Analyses of Fig. 3.11 have revealed that three types of frequency offsets (0, 10.15 and 22.5 MHz) are used to compare the minimum separation distances in the four environments.

It is important to mention that the FSS frequency carrier was simulated to be fixed at 4000 MHz and the WiMAX in the adjacent channel has been moved to suit the scenario requirements in the lower frequency. However, this could be the worst case compared to the upper frequency, and therefore the zero-guard band position is calculated as follows:

$$Zero\_Guard\_band\_Frequency = Fc_{FSS} - \left(Fc_{FSS}/2\right) \qquad (3.18)$$

where $Fc_{FSS}$ is the frequency carrier on the victim FSS and the zero-guard band is given by

$$Zero\_guard\_band = \frac{1}{2}\left(BW_{Interferer} + BW_{Victim}\right) \qquad (3.19)$$

Zero_guard_band is equal to 10.115 MHz for the 0.23 MHz FSS channel bandwidth.

By comparing Fig. 3.10 with the results presented in Fig. 3.11, it is observed that when FSS receiver bandwidth is 36 MHz, the possibilities of coexistence are increased due to separation distance reduction. However, the noise level of a victim FSS is improved to −143 dBW/36 MHz, and the adjacent channel scenario gave a wide separation between carriers. This indicates that the higher the difference between victim receiver bandwidth and that of the interferer, the less the effects of interference from the transmitter. This translates to brighter coexistence feasibilities. Alternatively, it is observed that when the frequency offset is more than half of the interferer bandwidth, the separation distance becomes significantly small.

Different deployment areas have an obvious impact on the results in Figs. 3.9, 3.10 and 3.11. However, considering three coexistence scenarios (CCI, zero-guard band and ACI), the simulation is done for dense urban, urban, suburban and rural areas, with respect to variable FSS bandwidths. Different antenna heights were used in the simulation process to evaluate the minimum separation distance with different clutter loss.

A remarkable change in the required separation distance is obtained in Figs 3.10 and 3.11 when different earth station antenna heights are used. Therefore, frequency coordination for different services to coexist in a same geographical area is apparently related to the antenna heights. It can also be noticed that increasing the antenna height from 1.8 m to 5 m (Fig. 3.11) changed the separation distance in the urban area. Similar observations were made in the case of the suburban area within the 12 MHz guard band. This clearly demonstrates the impact of the clutter loss.

By applying the interference calculation, a favourable assessment of the compatibility between the 20 MHz WiMAX and different fixed-satellite service receiver bandwidth is achieved.

## 3.7  Summary

A more straightforward approach to the protection ratio for future WiMAX 802.16e is derived and illustrated in order to achieve the frequency coordination with the FSS receiver. The use of MEASAT GSO downlink in the extended C-band (3.4–4.2 GHz) is used as a case study and has prompted this work to improve the mechanism of identifying the effective parameters between two base stations.

In the co-channel interference case, it is found that the adjacent channel selectivity reduction is unnecessary for the receiver when both services have the same bandwidth. However, if the interferer's bandwidth is larger than that of the victim, another factor must therefore be added to account for mask discrimination correction.

The ACIR is used for the adjacent channel interference scenario associated with the ACLR of the interferer (WiMAX 802.16e) and ACS of the victim (FSS receiver filter). The results have indicated that the required distance, as well as frequency separation, decreases as victim receiver bandwidth increases and vice versa.

Simulation results of MATLAB™ proved to be comparable to that of Visualyse Pro™ for the specified protection ratio. This includes the minimum required separation distance when frequency offsets in between carriers are equal to 28 MHz, 33 MHz and 40 MHz, respectively. Moreover, the proposed approach features a tractable as well as systematic workflow to the calculation of protection ratio. In the same time, it is applicable to a range of frequencies and bandwidths by simply calculating the required threshold degradation, ACLR and ACS.

It is worth mentioning that this method may also be used by spectrum regulation bodies to help in achieving best coexistence practices for a FSS receiver when coexist with an IMT-Advanced system.

## References

1. CEPT ECC Rep. 100, *Compatibility Studies in the Band 3400–3800 MHz Between Broadband Wireless Access (BWA) Systems and Other Services*, Bern, Switzerland, Feb 2007
2. L.F. Abdulrazak, Z.A. Shamsan, T.A. Rahman, Potential penalty distance between FSS receiver and FWA for Malaysia. Int. J. Publ. WSEAS Trans. Commun.. ISSN: 1109-2742 **7**(6), 637–646 (2008)
3. 3G Americas SM report, *3GPP Technology Approaches for Maximizing Fragmented Spectrum Allocations,* July 2009
4. Document AWF-3/17, *Assessment of Potential Interference Between Broadband Wireless Access (BWA) in 3.4–3.6 GHz Band and Fixed Satellite Service (FSS) in 3.4–4.2 GHz Band.* Office of the Telecommunications Authority (OFTA) Hong Kong, September 2006. http://www.esoa.net/v2/docs/public_cband/ESOA_CBand_APTReport.pdf
5. R.K. Hamid, F. Martin, L. Gérard, F. Eric, European harmonized technical conditions and band plans for broadband wireless access in the 790–862 MHz digital dividend spectrum. *Conference of European Post & Telecommunications Administrations.* Singapore, 6–9 April 2010. pp. 978–987

6. CEPT Rep. 19, *Report from CEPT to the European Commission in Response to the Mandate to Develop Least Restrictive Technical Conditions for Frequency Bands Addressed in the Context of WAPECS*, October 2008
7. Z.A. Shamsan, A.M. Al-hetar, T.A. Rahman, Spectrum sharing studies of IMT-advanced and FWA services under different clutter loss and channel bandwidths effects. Prog. In Electromagn. Res., PIER **87**, 331–344 (2008)
8. ITU-R WP 8F/TEMP 432 rev.2, *Working Document Towards a PND Report on Sharing Studies Between IMT-ADVANCED and the Fixed Satellite Service in the 3400–4200 and 4500–4800 MHz Bands*. ITU-R Working Party 8F, August 2006. http://www.itu.int/publ/R-REP-M.2109/en
9. IST-2003-507581 WINNER, *WINNER Spectrum Aspects: Methods for Efficient Sharing, Flexible Spectrum Use and Coexistence*. D6.1 v1.0, (2004)

# Chapter 4
# FSS Shielding and Antenna Discrimination as an Interference Mitigation Technique

## 4.1 Introduction

A tractable approach of dealing with interference issues imposed by the coexistence scenarios of FSS receiver and IMT-Advanced system is presented in this chapter. The IMT-Advanced system is represented by a 20 MHz bandwidth WiMAX IEEE802.16e base station. Coexistence analysis is done for CCI and ACI with guard bands equal to 0 MHz, 5 MHz and 12 MHz, respectively. The guard bands are used with shielding attenuation of 0 and 20 dB for each scenario. The proposed interference-to-noise ratio (*I/N*) is used as a prerequisite for a desensitization-proof receiver. A case study of using signal generator and VSAT unit as shielding materials is considered. The testing is performed in the anechoic chamber as well as outdoor, and deployment is designed to fulfil FSS signal receiving criteria. A set of key path loss parameters are calculated, followed by the computations of positive horizon angles imposed by losses over various terrains for different deployment areas. The antenna discrimination has been discussed alongside the shielding absorption coefficients of the suggested materials. Therefore, in this chapter shielding and other empirical measurements have been highlighted to reflect the pragmatism element of this study, in order to further enhance the response of the conventional theoretical effect.

## 4.2 Shielding Experiment and Tools

In the shielding experiment, the signal attenuation values by different proposed materials were first measured. These attenuation values are used as the effective factors in the interference mitigation between WiMAX and FSS.

© Springer International Publishing AG 2018
L.F. Abdulrazak, *Coexistence of IMT-Advanced Systems for Spectrum Sharing with FSS Receivers in C-Band and Extended C-Band*,
https://doi.org/10.1007/978-3-319-70588-0_4

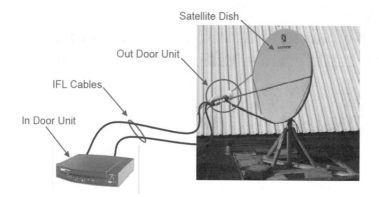

**Fig. 4.1** The FSS unit system

**Table 4.1** Fixed-satellite service specifications

| Specifications | Satellite terminal |
|---|---|
| Antenna diameter (m) | 1.8 |
| Antenna diagram | ITU RS.1245 |
| Frequency $F_c$ (MHz) | 4040 |
| Elevation angle | 74° |
| Azimuth | 263.7° |
| Height (m) | 1.8 |
| Received noise temperature | 114 K |
| Bandwidth (kHz) | 230 |
| Theoretical interference level (I) | −165 dBw/230 kHz |

One of the tools used for running the field test includes the VSAT unit, which receives the Internet signal at 4040 MHz. A harmful interference was applied on the FSS receiver using synthesized signal generator to generate interfered signal within the range 3700–4200 MHz. MEASAT-3 satellite network specifications and coverage are given in Appendix B, while the FSS Aspects and Installation are explained in Appendix F.

The FSS unit is installed to receive an Internet bandwidth (burstable to 256 kbps downlink and 9.6 kbps uplink) through MEASAT-3 geostationary satellite orbit. The FSS unit consists of dish antenna, C-band low noise block down converter, C-band 5 W block up converter and Indoor DW2000 Terminal. Figure 4.1 shows the FSS unit overall system receiver which has a carrier frequency of 4040 MHz. More details about the aspects are detailed in Appendix F.

The synthesized signal generator was used to generate an interference signal to assess the interference of 1 MHz bandwidth of broadband wireless signal. This is done to verify the effect of adjacent interference level as well as the in-band interference. Signal generator calibration was done to verify the error level in Appendix F. System specification is used to generate some of the most important parameters

**Table 4.2** Broadband wireless access specifications

| Specifications | Satellite terminal |
|---|---|
| Centre frequency of operation (MHz) | 4040 |
| Tx peak output power (dBm) | 20 |
| Channel bandwidth (MHz) | 1 |
| Antenna gain (dBi) | 6–10 |
| Antenna gain pattern | ITU-R F.1336 |
| Antenna height (m) | 2.2 |

used in the mathematical model in order to improve the coexistence between two services. Tables 4.1 and 4.2 show the FSS receiver and broadband wireless access specifications, respectively.

In Table 4.1 the MEASAT-3 satellite orbit position is 91.5° E, while the dish is located at latitude 1.558° N and longitude 103.6° E. The distance of the earth station to the satellite is 35,955 km. The signal delay is 239 ms for MEASAT-3.

## 4.3  Field Measurements

Radio propagation model needs to be verified with the shielding measurements to represent the mitigation technique effect on the received signal. The measurement procedure with the shielding technique is explained step by step as follows:

1. Antenna measurements: Horn antenna is used in the shielding experiment for transmitting site to represent the BWA sector. The radiation pattern measurements and return loss are given in the Appendix F. H-plan and V-plan are exhibited using the sigma plot software. The measured return loss of the horn antenna gave a good response for the frequency band 3700–4200 MHz.
2. Direct measurement of signal level: In order to measure the signal path attenuation at the VSAT unit, it is necessary to specify the signal strength in the free line-of-sight (LOS) condition. In that sense, empirical experiment has been conducted using the anechoic chamber to measure the free line-of-sight signal level, as shown in Fig. 4.2. The results are measured using the spectrum analyser, as in Fig. 4.3.

3. Concurrently, a broadband wireless access (BWA) signal generator is used as a WiMAX transmitter. Having set such a typical ambience, various types of metals are located in between the transmitter and the receiver as shown in Fig. 4.4.

The objective of step 3 is to measure the amount of signal penetration and power loss through different metallic materials. A far-field measurement has been used to evaluate the shielding within the anechoic chamber.

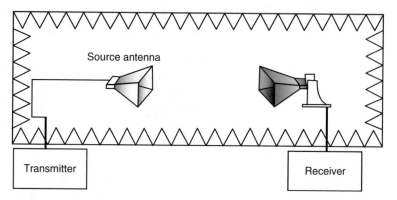

**Fig. 4.2** Free space signal level measurement

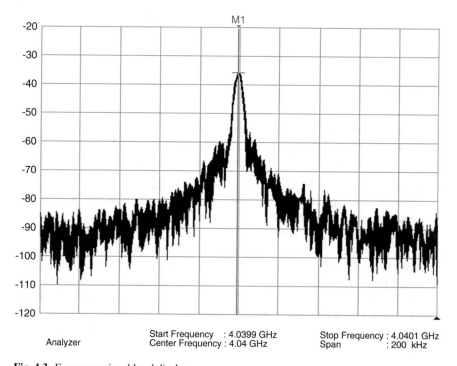

|  | Start Frequency : 4.0399 GHz | Stop Frequency : 4.0401 GHz |
| Analyzer | Center Frequency : 4.04 GHz | Span : 200 kHz |

**Fig. 4.3** Free space signal level display

4. Attenuation measurements for different materials: This is aimed at measuring the signal penetration through different materials in order to obtain the power loss through several barriers. The shielding is inserted between the transmitter-to-receiver path. Appendix F depicts the received power levels through different shielding. The results of attenuation obtained with different shielding materials are reported in Table 4.3.

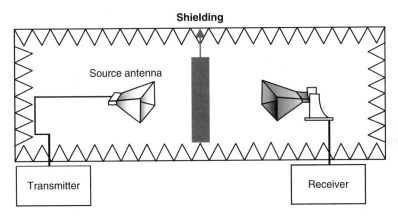

**Fig. 4.4** Different shielding materials between the signal generator and the spectrum analyser

**Table 4.3** Measurements of signal losses for 4040 MHz radio paths obstructed by common materials

| Material type | Loss (dB) |
|---|---|
| Aluminium shield (0.1 cm thickness) | 22.1 |
| Aluminium mesh wire shielding (0.2 cm wire spacing) | 20.9 |
| Copper shielding (0.1 cm thickness) | 24.5 |
| Copper mesh wire shielding (0.2 cm wire spacing) | 23.3 |
| Zinc shield (0.1 cm thickness) | 20 |

**Fig. 4.5** FSS shielding using zinc sheet material

The losses obtained are in the range of 20–22.1 dB for the materials used as shown in Table 4.3. Note that these materials have shown a diverse range of attenuation capabilities towards the radio signal. For cost-effective deployment, a zinc metal of 0.1 cm thickness is used for shielding the FSS as shown in Fig. 4.5. The mesh wire blocks the signal according to Faraday's law of induction [1]. This

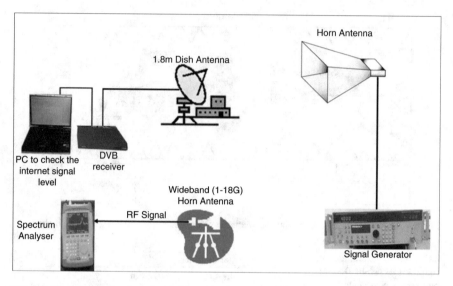

**Fig. 4.6** The interferer and victim tools used in the experiment

process ends when an electromagnetic field is no longer transmitted. The charge distribution on the mesh is known as induced charge distribution.

5. Intuitively, when the direction of the interferer is known, the FSS signal can be further detected by moving the antenna in such a way that the building is situated in between the interferer and the FSS antenna.

The study has also practically proved that best shielding condition occurs when the FSS receiver antenna is entirely shielded except for the top side. It should be uncovered and must be pointed to the satellite. Furthermore, the shield should be separated at least 1 m from the basement of antenna and 0.5 m higher than the antenna's body. It must also be grounded. If a shield is deployed in the direction of the satellite, the angle of elevation from the bottom of the antenna reflector to the top of the shield should be about 5 degrees less than the satellite elevation.

6. A co-channel and adjacent channel interference scenarios are identified and measured for the system parameters which have already been specified; Fig. 4.6 shows the interferer and victim tools used in the experiment.

Figure 4.7 shows the FSS antenna installed in one position and subjected to interference from all sights, for experimental purpose.

For all positions in Fig. 4.7, the received signals by FSS were 0 kb/s, while interfered signals were varying between 3800 MHz and 4100 MHz. By reducing the transmitting power level using power attenuator, the interference is reduced up to −125 dBm, while no interference is received by the FSS receiver.

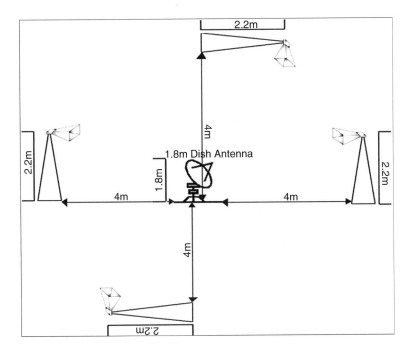

**Fig. 4.7** CCI and ACI scenario measurements

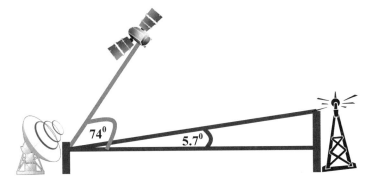

**Fig. 4.8** Determination of receiver gain using the elevation angle difference

To obtain the minimum separation distance using the deterministic calculations in the CCI scenario, the FSS elevation angle should be calculated according to the interferer position as illustrated in Fig. 4.8.

As shown in Fig. 4.8, the FSS off access angle can be determined as follow:

$$\alpha = 74^\circ - \tan^{-1}\frac{0.4}{4} = 68^\circ \tag{4.1}$$

**Fig. 4.9** Zinc sheet used in the FSS unit shielding experiment

This value will lead to −10 dB as an FSS antenna gain towards the interference source as mentioned in Eq. (3.3).

## 4.4   Fixed-Satellite Service Receiver Shielding

A 1.8 m VSAT unit is used to receive Internet signal through MEASAT-3 satellite on licensed band at 4040 MHz frequency carrier. On the interferer side, a BWA synthesized signal generator is used to generate an interfering signal, which ranges from 3400 to 4200 MHz. This frequency range covers both cases of CCI and ACI.

The FSS unit (256 kbps downlink and 9.6 kbps uplink) comprises of the dish antenna, C-band low noise block down converter, C-band 5 W block up converter and Indoor DW2000 Terminal. A horn antenna is used as an alternative to the directional antenna of WiMAX base station. Figure 4.9 shows the measurement site.

In order to shield the FSS unit, a 0.1 cm thick zinc sheet is selected for cost-effectiveness. A 20 dBm signal with a bandwidth of 1 MHz was generated and broadcasted in the direction of FSS receiver. The frequency of the interferer was varied from 3800 to 4100 MHz, which resulted in 0 kb/s downlink signal in the FSS receiver. However, when the transmitter power of interferer is reduced by 1 dBm, significant decrease in interference was observed. In order to have a minimum separation distance required for the coexistence in CCI scenario, the deterministic calculation is given by

$$20\log(d) = \mathrm{EIRP}\left(-65\mathrm{dBw}\right) - I\left(-165\mathrm{dBW} / 0.23\mathrm{MHz}\right)$$
$$+ G_{\mathrm{r}}\left(-10\right) - 92.44 - 20\log f\left(4.02\mathrm{GHz}\right); d = 0.187\,\mathrm{km} \qquad (4.2)$$

**Table 4.4** Effect of FWA signal on the FSS carrier with and without shielding

| Parameter | | 20 dBm signal generator | | | |
| | | Without shielding | | With 0.1 cm Zinc | |
| | | Before carrier | After carrier | Before carrier | After carrier |
|---|---|---|---|---|---|
| ACS (dB) | 5 MHz offset | 15 | 50 | 35 | 70 |
| | 10 MHz offset | 31 | 60 | 61 | 105 |
| | 15 MHz offset | 55 | 70 | 0 | 0 |

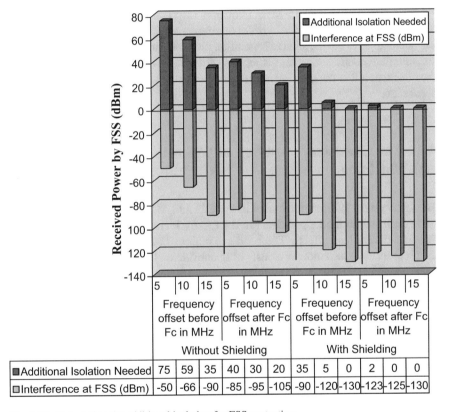

| | | | Without Shielding | | | | | | With Shielding | | | |
|---|---|---|---|---|---|---|---|---|---|---|---|---|
| | | | Frequency offset before Fc in MHz | | | Frequency offset after Fc in MHz | | | Frequency offset before Fc in MHz | | | Frequency offset after Fc in MHz |
| | | 5 | 10 | 15 | 5 | 10 | 15 | 5 | 10 | 15 | 5 | 10 | 15 |
| ■ Additional Isolation Needed | 75 | 59 | 35 | 40 | 30 | 20 | 35 | 5 | 0 | 2 | 0 | 0 |
| □ Interference at FSS (dBm) | -50 | -66 | -90 | -85 | -95 | -105 | -90 | -120 | -130 | -123 | -125 | -130 |

**Fig. 4.10** Calculating the additional isolation for FSS protection

where $d$ is the separation distance in km, EIRP is the effective isotropic radiation power of the interferer, $I$ is the interference level, $G_r$ is the received gain and $f$ is the receiving frequency of FSS. A 0.187 km is a large separation distance for a small transmitted power like 20 dBm. Therefore, the experiment has shown that coexistence scenarios based on co-channel sharing are almost practically impossible. The analysed interfered signal collected in Table 4.4 was used to ensure the wave propagation attenuation after and before the FSS frequency carrier.

As clearly shown in Table 4.4, the higher the transmission frequency, the higher is the propagation losses; and reducing the transmitted power corresponds to a reduced ability to penetrate the walls. Therefore, the effects of interference, with or without shielding, at different frequency offsets are summarized in Fig. 4.10. The threshold value is defined at −125 dBm (see Appendix F).

Figure 4.10 clearly shows that signal ability to interfere is bigger when it has less value than FSS frequency carrier and vice versa. In addition, signal attenuation is higher when shielding technique is used compared to the signal attenuation before the shielding. It is also shown in the figure that coexistence is achieved with 20 dB shielding attenuation and 15 MHz frequency offset. It can be therefore concluded that it is possible to reduce the harmful interference to 10% by increasing the shielding attenuation to 20 dB, and the separation distance can be reduced to 1% for 40 dB shielding attenuation. Careful inspection of the values shown in Fig. 4.10 leads to the conclusion that the coexistence under co-channel sharing is intolerable without separation.

## 4.5  The Coexistence Analysis of FSS with WiMAX Using Shielding

The intersystem interference model in the CCI and ACI depends essentially on the wave propagation model presented in Eq. (3.16). It consists of free space propagation and clutter loss effects. The band separations between different channels for different services have not previously been considered in the spectrum-sharing studies reviewed in Chap. 2. In addition, the deployment area nature (dense urban, urban, suburban and rural areas) should reflect different deployment possibilities according to local clutter loss.

According to the shielding experiment at CCI scenario, the zero-guard band and guard band separation channel are simulated to represent the interference scenarios. A minimum separation in two dimensions (frequency and distance) for different deployment areas with and without using the shielding technique has been studied. Different system scenarios have been used to improve the simulation results by considering different shielding attenuation values. The ACLR and ACS values have been considered for the transmitter and receiver, respectively.

Clearly in any coexistence analysis, a minimum separation in terms of frequency and distance is always sought-after and, preferably, with $(I/N) = -10$ dB under different deployment conditions, with or without shielding technique.

Firstly, the interfering signal shares the same band with the victim FSS receiver and thus separation distance is desired. Secondly, the interfering signal is contiguous to the victim band and finally when a guard band is in between the bands in question. The worst case of sharing between WiMAX and FSS receiver is simulated when both the interfering and victim antennas are opposite-tower-mounted and facing each other. The WiMAX used in the study is shown in Table 4.5. The FSS

**Table 4.5**  WiMAX specifications

| WiMAX systems parameters | |
| --- | --- |
| Centre frequency of operation (MHz) | 4040 |
| Channel bandwidth (MHz) | 20 |
| Base station transmitted power (dBm) | 43 |
| Spectral emissions mask requirements | ETSI-EN301021Type G |
| Base station antenna gain (dBi) | 18 |
| Base station antenna height (m) | Up to 30 |

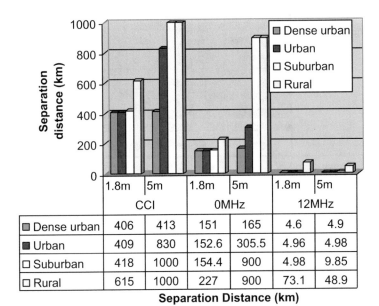

| | 1.8m CCI | 5m CCI | 1.8m 0MHz | 5m 0MHz | 1.8m 12MHz | 5m 12MHz |
| --- | --- | --- | --- | --- | --- | --- |
| ▣ Dense urban | 406 | 413 | 151 | 165 | 4.6 | 4.9 |
| ■ Urban | 409 | 830 | 152.6 | 305.5 | 4.96 | 4.98 |
| □ Suburban | 418 | 1000 | 154.4 | 900 | 4.98 | 9.85 |
| □ Rural | 615 | 1000 | 227 | 900 | 73.1 | 48.9 |

Separation Distance (km)

**Fig. 4.11**  The separation distance between WiMAX and FSS when FSS bandwidth is 0.23 MHz for four deployment areas for CCI, zero-guard band and 12 MHz guard band with 20 dB shielding attenuation

parameters mentioned in Table 3.4 are used for this simulation with 4040 MHz as a frequency carrier.

The WiMAX parameters in Table 4.5 reflect the fact that a fixed position has been used for the FSS receiver, while a movable type of WiMAX base station parameters is apparently shown. These parameters are used in the proposed propagation model of Eq. (3.12) to assess the interference between WiMAX and FSS in terms of minimum separation distances. An FSS antenna of variable heights (1.8 m and 5 m) has been used to emphasize that positioning the FSS receiver onto the ground can effectively reduce the separation as earlier shown in Figs. 3.10 and 4.11, respectively.

Figure 3.11 has earlier shown the unshielded scenarios of locating the FSS receiver without being protected. A 0 dB shielding was the input to Eq. (3.16) for

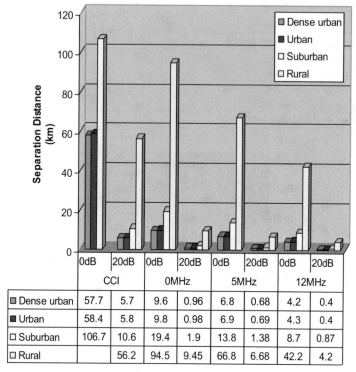

| | CCI | | 0MHz | | 5MHz | | 12MHz | |
|---|---|---|---|---|---|---|---|---|
| | 0dB | 20dB | 0dB | 20dB | 0dB | 20dB | 0dB | 20dB |
| ▣ Dense urban | 57.7 | 5.7 | 9.6 | 0.96 | 6.8 | 0.68 | 4.2 | 0.4 |
| ■ Urban | 58.4 | 5.8 | 9.8 | 0.98 | 6.9 | 0.69 | 4.3 | 0.4 |
| ☐ Suburban | 106.7 | 10.6 | 19.4 | 1.9 | 13.8 | 1.38 | 8.7 | 0.87 |
| ☐ Rural | | 56.2 | 94.5 | 9.45 | 66.8 | 6.68 | 42.2 | 4.2 |

**Separation Distance (km)**

**Fig. 4.12** The separation distance between WiMAX and FSS when FSS bandwidth is 36 MHz in the four deployment areas for CCI, zero-guard band and 12 MHz guard band and 40 MHz with 0 and 20 dB shielding attenuation

FSS bandwidth of 0.23 MHz, so that the minimum required separation distance between the two services is acquired for the CCI scenario (for 0 and 12 MHz guard band). A minimum separation distance between the two services was achieved using a dense urban environment as the deployment area, while the worst case of interference occurs when a rural area was used for deployment. A minimum separation distance without mitigation technique was obtained when 12 MHz was used as a guard band in urban and dense urban area.

In Fig. 4.11, the 20 dB shielding mitigation technique is used to reduce the separation distance between the two services, which corresponds to the reduction of 10% of the original distance (Refer to Fig. 3.14).

The reduced separation obtained in Figs. 3.10 and 4.11 (with the insertion of 12 MHz guard band and 20 dB Shielding) is not sufficient for practical deployment of the future communications systems. Obviously, the 36 MHz channel bandwidth of FSS will only need a small separation distance due to the reduced receiver susceptibility towards interference. Therefore, a minimum separation distance is calculated for 36 MHz FSS bandwidth when shielding attenuations are 0 and 20 dB, respectively. Figure 4.12 shows the results for 36 MHz FSS (1.8 m) bandwidth when $\Delta f = 0$ (CCI), 28 (zero-guard band), 33 (12 MHz guard band) and 40 MHz for the four deployment areas.

From Fig. 4.12, it is noticed that separation distance reduced to 0.4 km when 12 MHz is used as guard band (with a 20 dB as a shielding attenuation in dense urban area deployment). However, coexistence in the CCI scenario is still difficult due to large separation distance required. Accordingly, the values of separations in the rural areas can be neglected in order to keep a good resolution for other values in Fig. 4.12.

The worst case of interference was the first considered, when IMT-Advanced base station is considered without using any mitigation technique. Thereafter, protecting the FSS was the first target in order to study the effect of different elevation angles according to the user position and signal attenuation through different shielding materials. Since base station to base station is the main scenario of interference, complete analyses on the antenna discrimination effect should be done by using the smart antenna, being a suggested technology for next generation of mobile communication.

## 4.6   Antenna Discrimination Impact

An antenna discrimination loss (ADL) is the difference in azimuth between the interferer antenna direction and the victim receiving direction. Thus, pointing the beams of antenna victim and interferer is not aligned on each other, and it could lead

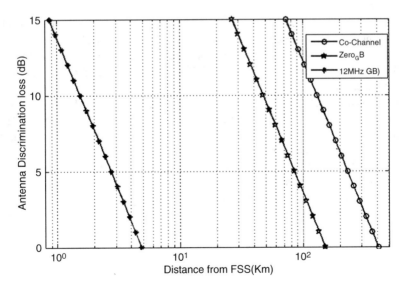

**Fig. 4.13** Antenna discrimination loss and minimum separation distance when FSS bandwidth is 0.23 MHz in dense urban area with 20 dB shielding attenuation

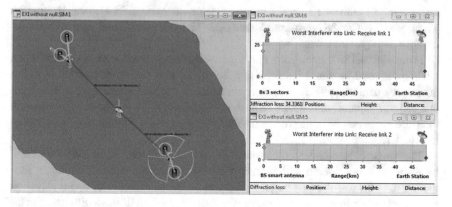

**Fig. 4.14** ADL scenario where the first link is between the FSS and three sector antennas and the second link is between the FSS and smart antenna system for ADL

to degradation in the interferer gain towards the victim. In order to highlight this issue, the separation distance results obtained in Fig. 4.11 for a dense urban area deployment are incorporated in the ADL simulation. Figure 4.13 shows the varying values of minimum separation distance using ADL in the range of 0–15 dB of the CCI, with 0 and 12 MHz guard band separation.

Definitely, the ADL technique proposes another mitigation technique which supports the smart antenna technology. However, a scenario of intersystem interference is also considered in order to compare the effect of three sectored terrestrial base station and electrically shifted beam base station on the FSS earth station. Both systems can have the same parameters in terms of power, coverage, antenna height and number of users.

In the cases of co-channel coexistence, zero-guard band and adjacent channel, a 15 dB antenna discrimination loss can decrease the physical separation from 406 km, 165 km and 49 km to 72 km, 26.8 km and 0.86 km, respectively, for 20 MHz WiMAX channel bandwidth and 1.8 m FSS height with 20 dB shielding protection in dense urban area.

A scenario was built using Visualyse to show the result of deploying FSS earth station in between two WiMAX base stations. The first WiMAX base station is three sectors supporting full coverage, whereas the second base station is transmitting signal based on switched beam smart antenna system coverage. Both base stations can support two mobile users moving around their base station following Monte Carlo distribution, as shown in Fig. 4.14.

Figure 4.14 shows that both base stations are deployed on an equally separated distance. The FSS unit uses the same signal propagation model and environment in order to determine its effect on the $I/N$ of the FSS. Statistical plot results are collected for comparing the $I/N$ received by FSS for 1 min duration as shown in Fig. 4.15.

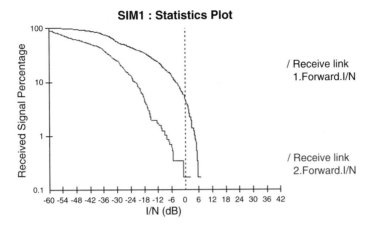

**Fig. 4.15** Long-term interference between two links

Fig. 4.15 shows that long-term interference occurs on the FSS receiver faster when the first link is functional, while for the second link interference-to-noise ratio has slow movement towards the protection ratio of $-10$ dB.

## 4.7 Summary

The proposed shielding technique proves resilient in the presence of interference (with or without the guard band). The technique may thus be considered as a viable alternative to many other commercial-of-the-shelf (COTS) mitigation solutions. The shielding offers both cost-effectiveness and an inherently high attenuation (20 dB). Furthermore, a prediction method that takes into account the bandwidth correction factor has been developed. This chapter has also exploited the propagation effects due to unwanted emission of WiMAX that falls within the FSS receiver range using variable antenna height.

When using a shielding attenuation of 20 dB, a separation distance can be substantially reduced to 10%. Consequently, the separation distance can be reduced to 1% for the 40 dB shielding attenuation. This method can be applied to other satellite systems, because different channel bandwidths were simulated for the victim FSS receiver. The simulation results have shown that both the interference and separation distance decrease with an increasing channel bandwidth.

Co-channel interference scenario in the rural area is the most difficult compared to other scenarios. However, it requires a long coordination distance in the range of 6150 km to 86 km without shielding effect for a 0.23 MHz and 36 MHz FSS channel bandwidths, respectively, given an FSS antenna height of 1.8 m. By adding 40 dB shielding attenuation, the coordination distance will correspondingly be reduced to 61 km and 0.86 km. These are the highest reduction that could be achieved without guard separation. These findings emphasize that the shielding technique can significantly improve the FSS immunity against the interference as well as the signal reception via FSS. However, adjacent channel interference scenario with frequency offsets from the carrier of 12 MHz in dense urban area shows the best coexistence scenario with 40 dB shielding attenuation. For instance, it needs 0.49 km and 0.04 km geographical separation for 0.23 MHz and 36 MHz FSS channel bandwidths, respectively, when FSS antenna height is 1.8 m. This indicates that the dense urban area is the best area for coexistence and intersystem interference coordination.

From the deployment standpoint, different areas are considered, and it is shown that the dense urban type of environment is the most convenient type for successful coexistence scenarios, whereas the rural one is the worst for frequency sharing and coordination in the same band. From the shielding perspective, it is worth mentioning that this technique is applicable to any antenna size at various heights. Added advantage can be achieved when antennas are installed at heights equal or less than the heights of surrounding obstacles and/or clutters. For the ADL, it is concluded that other mitigation techniques should be researched to enhance the coexistence between the two services by reducing the separation distance.

# Reference

1. J.L. Drewniak, *Coupling through the Magnetic Field--Faraday's Law* (EMC Laboratory, University of Missouri-Rolla, 2010), pp. 1–9. http://www.emcs.org/edu/ExperII/MagField Coupling.pdf

# Chapter 5
# I-MUSIC Algorithm and Fixed Null Insertion

## 5.1 Introduction

A hybrid approach to solve the problem of coexistence between the fixed-satellite services (FSS) receiver and WiMAX base stations (IEEE802.16e) in the 3400–4200 MHz band is proposed in this chapter. The hybrid part stems from a blend of two popular algorithms used in adaptive breamforming, namely, multiple signal classification (MUSIC) and least mean square (LMS). It is used to steer the beam in the direction of WiMAX's users while nulling those emitted towards the victim's FSS earth station. The problem of high resolution and accurate direction of arrival detection using proposed Improved MUSIC (I-MUSIC) algorithm is highlighted. I-MUSIC prevents the high heavy complexity of currently available methods earlier mentioned in the literature. The amount of interference appears on the VSAT receiver because the WiMAX base station (BS) is calculated after using the interference mitigation measures in both co-channel interference (CCI) and adjacent channel interference (ACI) scenarios. In order to further validate the proposed technique and its applicability to different operating environments, certain situations are considered. These include various shielding attenuations, FSS bandwidths, guard bands and deployment areas which are considered in line with the minimum separation distance.

## 5.2 Problem Identification

The major focus of this chapter is improving the adaptive antenna type due to its inherent sort of built-in intelligence over the switched array type. The processing unit of the adaptive array system is located inside the WiMAX BS. It uses the

© Springer International Publishing AG 2018
L.F. Abdulrazak, *Coexistence of IMT-Advanced Systems for Spectrum Sharing with FSS Receivers in C-Band and Extended C-Band*,
https://doi.org/10.1007/978-3-319-70588-0_5

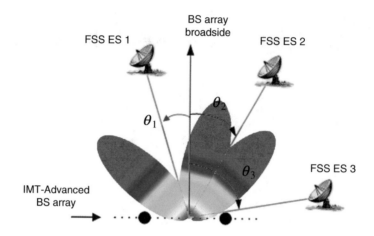

**Fig. 5.1**  Adaptive array scheme to avoid the interference in the FSS ES

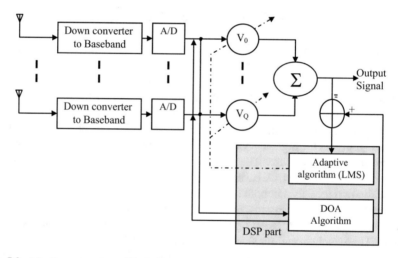

**Fig. 5.2**  Adaptive array system block diagram

predefined position of the victim FSS receiver to steer the main beam towards the desired WiMAX user and block those pointed at the victim FSS receiver [1–7].

As the FSS receiver emits no signals, it is difficult to inform the WiMAX BS about the exact location of the FSS receiver, unless these coordinates are manually included in the WiMAX BS. This is much superior to the performance of a switched-beam system, as shown in Fig. 5.1.

Figure 5.1 shows that the adaptive beam system can position the desired main lobe to the user direction and also exhibits the ability to fully null the interference in the FSS ES direction.

Figure 5.2 shows the block diagram of the proposed system. Clearly, the output of the digitization process (i.e. signal of interest (SOI)) after being down-converted is fed to the direction-of-arrival (DOA) unit for further processing.

Figure 5.2 shows that the outputs of the antennas are linearly combined after being weighted, while the weights are computed using LMS algorithm based on minimum square error (MSE). The LMS algorithm is used to estimate the pattern from the received signal by minimizing the error between the reference signal and beamformer output. Finally, the optimum weight can be iteratively found using the LMS algorithm.

## 5.3   Improving MUSIC Algorithm for High-Resolution DOA Detection

MUSIC algorithm is based on eigenvector decomposition. It provides a good resolution and sharp peaks with the ability to add nulls. As mentioned in Chap. 2, improving the resolution degree of MUSIC algorithm is still a challenge especially for multiuser with small angles. This section focuses on improving the MUSIC algorithm by developing the power spectrum substitution.

### 5.3.1   Steering Vector

Assume $x(t)$ is the vector process of the complex envelopes of the signals at the output of an array of $M$ narrow band identical sensors. A specific number of rays $P$ transmitted to $M$ with half-wave length spacing are denoted as $P$-dimensional source vector. The received base-band signal can be expressed as

$$x(t) = \sum_{P=1}^{P} s_p \mathbf{a}(\theta_P) + \mathbf{n}(t) = \mathbf{A}(\Theta)\mathbf{S}(t) + \mathbf{n}(t) \tag{5.1}$$

where $\mathbf{S}(t) = [s_1(t), \ldots, s_p(t)]^T$ is the data inside of $P$th array, $(.)^T$ denote the transpose function and $\mathbf{A}(\Theta) = [\mathbf{a}(\theta_1), \ldots, \mathbf{a}(\theta_P)]$ is the array direction vector for DOA $\theta_P$ of the $P$th ray. It is assumed that all sources emit a common pulse shape.

By considering only one sample by deploying $b$ times symbol rate during $C$ periods, then the collected samples can be obtained by $MC \times b$ matrix:

$$MC \times b = \begin{bmatrix} x(0) & x(1) & \cdots & x(C-1) \\ x\left(\dfrac{1}{b}\right) & x\left(1+\dfrac{1}{b}\right) & \cdots & x\left(C-1+\dfrac{1}{b}\right) \\ \cdot & \cdot & & \cdot \\ \cdot & \cdot & \cdots & \cdot \\ \cdot & \cdot & & \cdot \\ x\left(1-\dfrac{1}{b}\right) & x\left(2-\dfrac{1}{b}\right) & \cdots & x\left(C-\dfrac{1}{b}\right) \end{bmatrix} = X \tag{5.2}$$

By taking into account the Fourier transform of the received sampled output, the covariance matrix $\mathbf{R}_x$ can be obtained, which is emitted by $\overline{\mathbf{X}}\overline{\mathbf{X}}^H / C$, as follows:

$$\hat{\mathbf{R}}_x = E\left[\mathbf{x}(t)\mathbf{x}(t)^H\right] = \mathbf{E}_s\mathbf{D}_s\mathbf{E}_s^H + \mathbf{E}_n\mathbf{D}_n\mathbf{E}_n^H$$

$$= \mathbf{S}\mathbf{R}_{xx}\mathbf{S}^* + \sigma^2\mathbf{I}_Q \qquad (5.3)$$

$$= \begin{bmatrix} \mathbf{R}_{x0} & \cdot & \cdot & \cdot & \mathbf{R}_{xP} \\ & \cdot & & & \cdot \\ & & \cdot & & \\ & & & \cdot & \\ \mathbf{R}_{xP} & \cdot & \cdot & \cdot & \mathbf{R}_{x0} \end{bmatrix}$$

where $E$ is the statistical expectation process, $\mathbf{E}_n$ represents subspace noise component, $(.)^H$ denote the conjugate transpose function, $(.)^*$ denote the complex conjugation, $\sigma^2$ is the noise standard deviation, $\mathbf{I}_Q$ is a $Q \times Q$ identify matrix, $\mathbf{D}_s$ is a $P \times P$ diagonal matrix including the largest P eigenvalues and $\mathbf{D}_n$ is a diagonal matrix for the smallest eigenvalues. Therefore, $\mathbf{E}_s$ will respond to the largest $P$ eigenvalues of $\hat{\mathbf{R}}_x$ . $\mathbf{E}_n$ represents matrix including the rest of eigenvectors. Then, the steering vector which corresponds to the values in angles is given by

$$\mathbf{A} = e^{\frac{J2\pi d}{\lambda}\cos(\theta_P)\times\frac{\pi}{180}\times\{0...M-1\}} \qquad (5.4)$$

where $\lambda$ is a wavelength, $d$ is the space between two elements and $d = 0.5\ \lambda$.

## 5.3.2   Improved Music Algorithm

The complexity of eigenvalue decomposition is the main difficulty when processing MUSIC algorithm in real time due to its computationally intensive requirements. Thus, the main objective is to reproduce the MUSIC spectrum in order to improve the overall resolution. According to Eq. (5.3), the MUSIC spectrum function can be constructed and computed for 180°, as follows:

$$P_{\text{MUSIC}}(\theta) = \frac{1}{\mathbf{a}^H(\theta)\mathbf{E}_n\mathbf{E}_n^H\mathbf{a}(\theta)} \qquad (5.5)$$

where $\mathbf{a}(\theta)$ is

$$\mathbf{a}(\theta) = \begin{bmatrix} 1 & e^{\frac{-j2\pi d}{\lambda}\cos\left(\frac{a\pi}{180}\right)} & \cdot & \cdot & \cdot & e^{\frac{-j2\pi d}{\lambda}(M-1)\cos\left(\frac{a\pi}{180}\right)} \end{bmatrix} \qquad (5.6)$$

As found in the literature, all previous improvements on MUSIC algorithm require an exhaustive searching, which is inefficient due to high computational cost. I-MUSIC algorithm is presented in this section, which is qualified for high resolution by using one dimension of searching.

I-MUSIC algorithm is based on the fact that covariance of steering vector is actually not necessary because it can increase the power function. However, at the same time, it can reduce the accuracy to reach the mean. In order to increase the resolution, noise subspace correlation matrix will be correlated with mode vector to represent the power from the subspace. Finally, the new proposed I-MUSIC spectrum is

$$\text{PMU}(\theta) = \frac{1}{\mathbf{a}^H(\theta)\mathbf{E}_n\mathbf{E}_n^H \left| 1 \quad 0 \quad . \quad . \quad . \quad (M-1) \right|} \tag{5.7}$$

Even though the expression of Eq. (5.7) is rather similar to that of Eq. (5.5), which is used for estimating DOA of ULA sensors, the proposed method is dedicated in order to have extra high resolution than what is proposed in [8]. Moreover, the method can be used under a lower SNR with a quite impressive resolution. It should be emphasized that the proposed method has better estimation performance than MUSIC-Like and MI-MUSIC which have been proposed in [9]. I-MUSIC is recognized as a generalization of MUSIC. The next expressions are adopted for estimating the DOA without noise. The signal subspace can be denoted as

$$\mathbf{E}_s = \mathbf{Y}\begin{bmatrix} \mathbf{A}_\theta\mathbf{D}_1 \\ . \\ . \\ . \\ \mathbf{A}_\theta\mathbf{D}_b \end{bmatrix} = \mathbf{Y}\begin{bmatrix} \mathbf{A}_\theta \\ . \\ . \\ . \\ \mathbf{A}_\theta\Phi^{b-1} \end{bmatrix} = \Lambda\mathbf{Y} = \left[\mathbf{a}(\theta_1),_{,.} \quad . \quad . \quad .,_{,}\mathbf{a}(\theta_p)\right]\mathbf{Y} \tag{5.8}$$

where $\mathbf{Y}$ is a $P \times P$ matrix and $\Phi$ is the rotational matrix and equal to diagonal of $\left[\exp(-j2\pi/b). \quad . \quad .\exp(-j2\pi/b_p)\right]$. $\hat{\mathbf{E}}_s$ is estimator of $\mathbf{E}_s$ that can be attained by scanning for the deepest $\mathbf{V}$ minimum as

$$P = a(\theta)^H\mathbf{Q}\left|10 \quad . \quad . \quad .(M-1)\right| \tag{5.9}$$

where $\mathbf{Q} = \mathbf{I}_M^H\Pi_{\hat{\mathbf{E}}_s}^{90^0}\mathbf{I}_M$. Equation (5.10) is the problem of quadratic optimization. Then, the constraint of $e_1^Y a(\theta) = 1$ is considered, where $e_1 = \begin{bmatrix} 1 & , & 0 & ,. & . & . & ,0 \end{bmatrix}^Y \in \mathbf{R}^{M\times 1}$ has been added to estimate the trivial solution $a(\theta) = \mathbf{0}_M$.

The optimized problem can be reconstructed with the minimum variance solution. By scanning from $0^\circ$ to $180^\circ$, it is found that the largest peak $V$ corresponds to the DOA. It can be represented with narrower bands. The major steps for I-MUSIC are explained as follows:

**Fig. 5.3** Computational
flow of I-MUSIC method

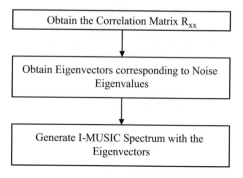

Step1: Searching for the covariance matrix of the received signal, in order to measure the changes of the variables.

Step2: Performing the eigen-decomposition associated with the covariance in a square matrix.

Step 3: Searching for the largest peaks $P's$ of the $b(\theta)^{-1}$ elements at the required DOA signals.

In contrast, the I-MUSIC algorithm can have the same computational load to MUSIC and lower complexity than MUSIC-Like and MI-MUSIC.

Accordingly, a MATLAB computer program was designed and used to estimate the direction of arrival. In addition, this program can adjust the angle of the FSS manually in order to avoid the interference (see Appendix D).

The computational flow chart of DOA estimation by the spectral MUSIC algorithm is illustrated in Fig. 5.3.

### 5.3.3   I-MUSIC Validation

A random binary phase shift keying (BPSK) signal is generated and accompanied by uncorrelated random noise of the arrival signals. It is used to assess the simulation performance based on a data matrix given in Eq. (5.3). It is generated and autocorrelated in order to get the covariance matrix. In the following simulation, it is assumed that there are four coherent rays arriving at the antenna array. Their DOAs are 75°, 90°, 105° and 120°. The number of estimated sources is based on Akaike's information criterion by calculating the root minimum square error

$$\text{RMSE} = \sqrt{\frac{1}{200}\sum_{n=1}^{200}(x-\bar{x})^2}$$ , where $x$ is the correct angle of arrival for $P$th array of

200 independent experiment and $\bar{x}$ is the perfect angle of arrival.

Figure 5.4 presents the angle estimation for I-MUSIC algorithm with number of sensors $M = 8$, $b = 10$, number of snapshot $C = 100$ and SNR = 15 dB. The spectrum peaks can be clearly observed at the detected angles, the evidence that the algorithm is performing well. Figure 5.5 shows the angle estimation performance with $M = 8$,

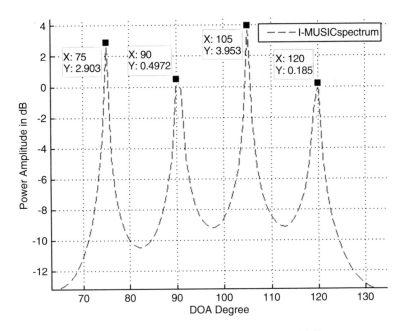

**Fig. 5.4** The angle estimation performance with I-MUSIC of SNR = 15 dB

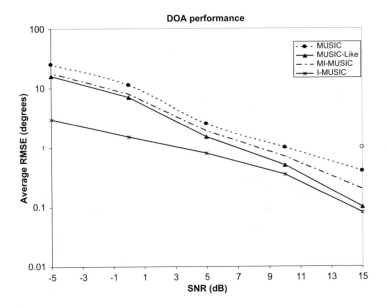

**Fig. 5.5** Angle resolution accuracy comparison with $C = 50$

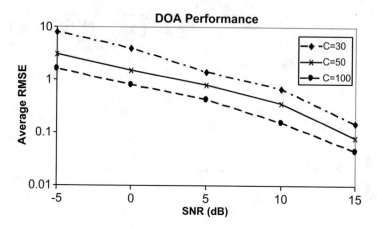

**Fig. 5.6** Angle estimation with different values of $C$

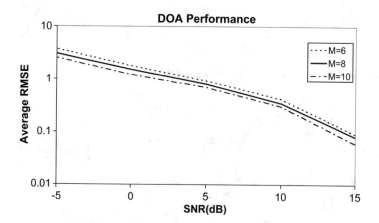

**Fig. 5.7** Angle estimation with $C = 50$ and different $M$

$b = 10$, $C = 50$, which the comparison is made for I-MUSIC algorithm with MUSIC, MI-MUSIC and MUSIC-Like methods. It is indicated in Fig. 5.6 that among the four algorithms considered, the I-MUSIC presents best resolution in the environment including noise and fading. The accuracy level of the angles is close to each other, which indicates that the highest amount of accuracy is for the new I-MUSIC. The lowest level of accuracy for normal MUSIC is due to natural processes of the mean calculation for this method.

Figure 5.6 depicts the algorithmic performance comparisons where I-MUSIC has been adopted. The simulation is shown for different $C$ (the same $M = 8$, $b = 10$ as Fig. 5.5). It is indicated that the performance of angle estimation becomes better when $C$ is increased. Figure 5.7 illustrates the angle estimation performance of I-MUSIC algorithm for different $M$. It is clearly shown that the estimation performance of I-MUSIC is gradually improved with the increasing number of antennas.

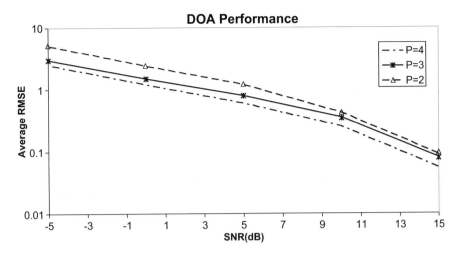

**Fig. 5.8** Angle estimation with $C = 50$ and different numbers of $P$

Multiple antennas improve the angle and delay estimation performance because of diversity gain. Figure 5.8 shows the algorithmic performance of I-MUSIC under different $P$ when $M = 8$, $b = 10$ and $C = 50$. From Fig. 5.8, it is concluded that the angle estimation performance levels are gradually decreased when the source number is increased.

In this section, the I-MUSIC algorithm is derived for blind estimation. It avoids the high computational process of other subspace-base algorithms. The I-MUSIC can have much better performance for angle estimation in contrast to MUSIC, MI-MUSIC and MUSIC-Like algorithms. This algorithm can also work well in other manifold arrays, and it can be expanded to other environments. The I-MUSIC algorithm can be regarded as a generalization of MUSIC.

## 5.4   Least Mean Square (LMS) Algorithm Implementation

LMS algorithm is used to estimate the gradient vector from acquired data. LMS incorporates an iterative procedure to correct the weight vector in the direction of the negative gradient vector which leads to the MSE [10]. An estimator of the gradient will be used as a substitute to the actual value of the gradient in order to avoid the cross-correlation and autocorrelation process. The iterative equation that updates the weight periodically is given by

$$\overline{V}_{k+1} = \overline{V}_k - \mu \nabla_k \xi \qquad (5.10)$$

where $\overline{V}$, $\mu$ and $\nabla_k \xi$ are the weight vector, step size and the gradient vector, respectively, where the step size is given by

$$0 < \mu < \frac{1}{\lambda_{max}} \tag{5.11}$$

where $\lambda$ max is the eigenvalue of the covariance matrix $R_{xx}$, when $X(k)$ is the input signals. It is clear that the performance can be evaluated by $\xi$. The performance function can be calculated using the following relation:

$$\xi = E\left[e^2(k)\right] = E\left[d^2(k)\right] - 2\bar{v}^T \bar{P} + \bar{v}^T R\bar{v} \tag{5.12}$$

where $R$ is the input autocorrelation matrix and $P$ is the cross-correlation vector between input signal $x(k)$ and error signal $d(k)$, respectively. They can be represented as follows:

$$R = E\left[\bar{x}(k)\bar{x}^T(k)\right] \tag{5.13}$$

$$\bar{P} = E\left[\bar{x}(k)\bar{d}^T(k)\right] \tag{5.14}$$

To find the gradient vector $\nabla_k \xi$, Wiener-Hopf equation should be applied, as follows:

$$\nabla_k \xi = 2R\bar{v} - 2\bar{P} \tag{5.15}$$

Finding the gradient vector simplifies the calculation in real-time applications. Computing the weights by LMS algorithm will be done as follows:

$$\nabla_k \xi = -2\bar{x}(k)\left[E\left(\bar{d}^T(k)\right) - E\left(\bar{x}^T(k)\bar{V}(k)\right)\right] \tag{5.16}$$

$$\bar{e}(k) = -\bar{x}^T(k)\bar{V}(k) + \bar{d}(k) \tag{5.17}$$

$$\bar{V}(k+1) = \bar{V}(k) + 2\mu.\bar{e}^T(k)\bar{x}(k) \tag{5.18}$$

where $V(k)$ is the weight vector, $x(k)$ is the input vector, $d(k)$ is the desired output and $V(k + 1)$ is the updated weight vector.

## 5.5 Implementation of Interference Mitigation Algorithm

Several steps need to be taken in order to implement the I-MUSIC. Firstly, the MUSIC algorithm is defined as a basis for the noise subspace and then determined by the peaks of the associated angles provided by the DOA estimation. Then, the MATLAB code of I-MUSIC algorithm is sampled by creating an array of steering vectors correspondent to the angles in the angles vector (see Appendix D). Next, a

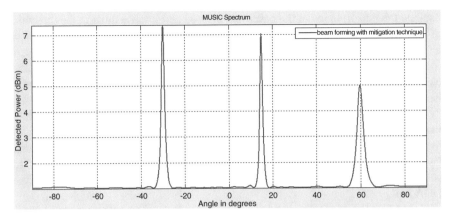

**Fig. 5.9** WiMAX BS radiation patterns ($Q$ = 3, DOE = 2), without interference mitigation technique

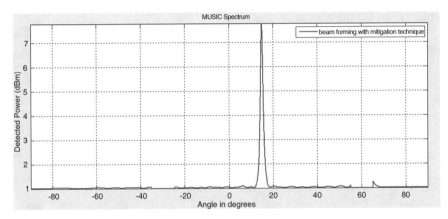

**Fig. 5.10** WiMAX BS radiation patterns ($Q$ = 3, DOE = 2), with interference mitigation technique

random binary phase shift keying (BPSK) signal is generated and accompanied by the uncorrelated random noise of the arrival signals. After that a data matrix ($x(t)$) as in Eq. (5.1) will be generated and autocorrelated ($R_{xx}$) to get the covariance matrix as in Eq. (5.3). Thereafter, the eigen-decomposition of covariance matrix is computed in order to find any arrival signal for the largest eigenvalues. The eigenvectors are then sorted in order to locate the signal eigenvectors before the noise eigenvectors. Finally, the signal eigenvectors and the noise eigenvectors will be defined.

By computing the steering vectors (which have corresponding values in angles) based on Eq. (5.6), I-MUSIC algorithm mathematical expression of Eq. (5.7) must be incorporated in the code in order to compute the angles for $180°$. Subsequently, a special expression makes the code more applicable when zeros are added in the spectrum to steer the beam at certain angles to represent the DOEs. In order to dem-

onstrate the performance of the described method and impose nulls in the direction
of the interfering signal, examples of I-MUSIC spectrum with $Q = 3$ and DOE $\hat{\theta} =$
2 (at $-30°$ and $60°$) and one-half wavelength spaced isotropic elements ($d$) are performed (using 500 data as snapshots of uniformly incoming power and eight antenna
elements). Figures 5.9 and 5.10 show MUSIC spectra for two DOEs when three
beams are received with only one beam being generated towards the WiMAX user.

For WiMAX BS to perform beam cancellation, the DOE needs to be manually
included with the local frequency administrator. As explained, the DOA is usually
included in the DSP part and can obtain the direction of SOI in a command figure.
By having that unit installed and properly set, an adaptive algorithm can be used to
steer the beam in the direction of users according to the DOA(s) provided by the
DOA unit.

A MATLAB code is designed to reflect the function of LMS algorithm in performing the beamforming functionality (see Appendix D). A linear array antenna of
isotropic elements is simulated by the code. When two nulls are identified at known
DOE and the $m$th weight vector is identified in the row vector, the plane wave
impinges on the antennas array at angle $\theta$.

Figure 5.11 shows an example of DOA, in which LMS unit is commanded to
steer the beam at $15°$ and locate the nulls at $-30°$ and $60°$. For an eight-element linear
array with inter-element spacing of $0.5\lambda$ between isotropic antennas, the LMS algorithm is used to obtain the normalized amplitude and phase coefficients. Resulting
patterns are collected and plotted as shown in Fig. 5.11 and Table 5.1.

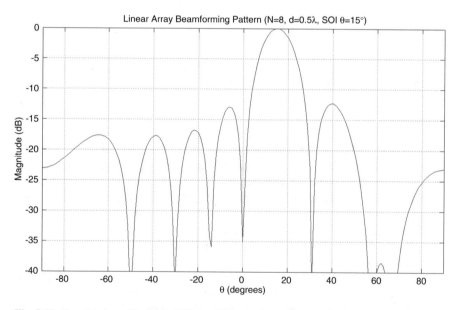

**Fig. 5.11** Forming the null towards FSS at $-30°$ and $60°$ and the SOI at $15°$

**Table 5.1** The amplitude and phase coefficients of an eight-element array using the LMS algorithm ($d = 0.5\lambda$, SOI = 15°, SNOI = −30° and 60°, $\mu = 0.001$, 245 iterations)

| Element number | LMS ($i = 245$) | | | |
| --- | --- | --- | --- | --- |
| | Weight value | Weight normalized | Phase value | Phase normalized |
| 1 | 0.109839 | 1.000000 | −3.203946 | 0.000000 |
| 2 | 0.124394 | 1.132509 | −38.153547 | 325.050400 |
| 3 | 0.142105 | 1.293762 | −95.733393 | 267.470554 |
| 4 | 0.115686 | 1.053234 | −146.22160 | 216.982339 |
| 5 | 0.114582 | 1.043180 | −179.25846 | 183.945485 |
| 6 | 0.142996 | 1.301869 | −231.30473 | 131.899215 |
| 7 | 0.124100 | 1.129835 | −287.09391 | 76.110032 |
| 8 | 0.108395 | 0.986858 | −322.995262 | 40.208685 |

Clearly, the radiation pattern of the simulated scenario abides by the requirements of power reduction in specific directions. It is also observed that by controlling the transmitted beam, it can reduce the overall gain in the victim direction.

## 5.6 The Coexistence Scenarios Between WiMAX and FSS

A viable interference mitigation technique is required for robust coordination policy between IMT-Advanced systems and FSS systems receivers. Since IMT-Advanced systems are not yet in existence, a Worldwide Interoperability for Microwave Access (WiMAX) 802.16e system is chosen to represent the physical configuration of IMT-Advanced to coexist with FSS receiver. Thus, interference power will be calculated at the FSS ES when WiMAX BS is operating with the proposed interference mitigation techniques.

Naturally, since different deployment environments have different distortion factors to the transmitted signal, ITU-R452.12 propagation model is used to calculate the clutter loss calculations of terrestrial communications [11]. It is assumed that dense urban, urban, suburban and rural areas are the operating environments for the WiMAX 802.16e. With the addition of minimum coupling loss, separation distance can be calculated from

$$20\log(d) = -I + P_{BS} + G_{ANT}\phi + G_{BF}\phi - 92.5 - 20\log(F) - A_h + G_{vs}(\alpha) \quad (5.19)$$

where $I$ is the received interference in dBW/MHz and EIRP$_{interferer}$ is the effective isotropic radiated power of the interferer (which consists of transmit power ($P_t$) and gain of transmitter ($G_{BS}$)). The transmitted gain is obtained by adding the $G_{ANT}\phi$ and $G_{BF}\phi$ together. The $G_{ANT}\phi$ is the conventional BS antenna pattern without interference mitigation techniques, and it is specified by

**Table 5.2** WiMAX system parameters

| Parameter | Value WiMAX |
|---|---|
| Centre frequency of operation (MHz) | 4000 |
| Channel bandwidth (MHz) | 20 |
| Base station transmitted power (dBm) | 43 |
| Spectral emissions mask requirements | ETSI-EN301021Type G |
| Base station antenna gain (dBi) | 18 |
| Base station antenna height (m) | 30 |

$$G_{ANT}\phi = G_{max} - \min\left[12\left(\frac{\phi}{\phi_{3dB}}\right)^2, A_m\right] \tag{5.20}$$

where $\phi$ is in the range of $-180°$ up to $180°$, $G_{max}$ is the maximum antenna gain (18 dBi), bandwidth at 3 dB is equal to $70°$ and the maximum attenuation $A_m$ is equal to 40 dB. $G_{BF}\phi$ is adaptive beamforming pattern generated by the null steering mitigation technique, and it is expressed as

$$G_{BF}(\phi) = 20\log_{10}\left|\frac{P_{Bs}}{element.numbers}\sum_m V_m^T a(\phi)\right| \tag{5.21}$$

The receiving gain of FSS station's off-axis antenna, $G_{vs}(\alpha)$, for a given off-axis is calculated to be $-10$ dB.

An IMT-Advanced system working frequency of 4040 MHz, with a bandwidth of 20 MHz in dense urban, urban, suburban and rural macrocell environments, is assumed to be able to support the proposed interference mitigation technique. IMT-Advanced BS parameters are presented in Table 5.2. The transmit power (EIRP) is 43 dBm and the maximum antenna gain is 13 dBi and the antenna height is 30 m.

With reference to the ITU-R Recommendation, FSS earth station parameters under consideration of sharing between the FSS and other services are shown in Table 3.4. The system occupies variance bandwidths of 230 kHz and 36 MHz assigned with a centre frequency of 4040 MHz. A dish-shaped directional antenna having a diameter of 1.8 m, height of 1.8 m and 5 m and a maximum antenna gain of 38 dBi is deployed. It should be kept in mind that FSS position is fixed and the WiMAX transmitter position is varying in order to achieve the coexistence.

## 5.7  The Assessment Results

Sharing scenarios categorized by co-channel band sharing and adjacent channel compatibility based on frequency domain have been simulated in order to obtain the best coexistence between two services when the null synthesized is applied. The

superiority of the proposed interference mitigation scheme is demonstrated by calculating the interference power in different clutter and shielding effects.

As mentioned earlier, shielding and null steering are intended to block the interfering signal, which originated from the WiMAX BS towards FSS receiver; it can
translate the interference into useful power towards the desired user.

The impact of fixed null synthesizing algorithm may be verified by having coexisted systems, deployed at different terrestrial areas with various antenna heights,
shielding and interference scenarios. Sharing scenarios are categorized into the CCI
and ACI depending on the frequency domain. If a flat service is considered, the
separation distance between FSS ES and WiMAX may reach up to 10,000 km and
6150 km for 5 m and 1.8 m FSS's antenna heights, respectively.

By considering the channelization plan for MEASAT downlink transponders, a
4 GHz central frequency channel is assumed. A 12 MHz guard band can be obtained
when a WiMAX BS of 20 MHz channel bandwidth replaces that of 36 MHz FSS
receiver's bandwidth.

When a fixed pattern synthesizer scheme is employed within the WiMAX BS
through the dense urban area, the separation distances are reduced to 123 m and
306 m for 1.8 m and 5 m FSS receiver's heights, respectively, as shown in Fig. 5.12.

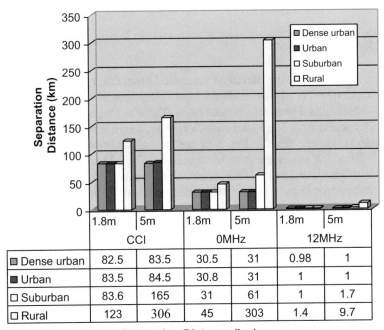

| | 1.8m | 5m | 1.8m | 5m | 1.8m | 5m |
|---|---|---|---|---|---|---|
| | CCI | | 0MHz | | 12MHz | |
| ■ Dense urban | 82.5 | 83.5 | 30.5 | 31 | 0.98 | 1 |
| ■ Urban | 83.5 | 84.5 | 30.8 | 31 | 1 | 1 |
| □ Suburban | 83.6 | 165 | 31 | 61 | 1 | 1.7 |
| □ Rural | 123 | 306 | 45 | 303 | 1.4 | 9.7 |

**Separation Distance (km)**

**Fig. 5.12** The separation distance between WiMAX and FSS with 0.23 MHz FSS bandwidth in
the four deployment areas for $\Delta f = 0$, 10.115 and 22.115 MHz with null synthesized method and
20 dB shielding

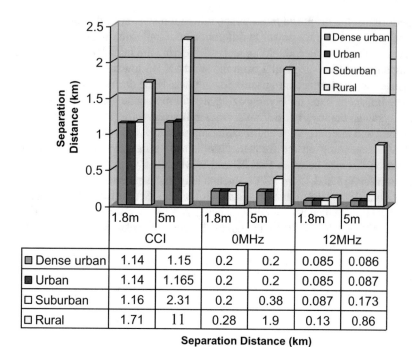

| | 1.8m | 5m | 1.8m | 5m | 1.8m | 5m |
|---|---|---|---|---|---|---|
| | CCI | | 0MHz | | 12MHz | |
| ◼ Dense urban | 1.14 | 1.15 | 0.2 | 0.2 | 0.085 | 0.086 |
| ◼ Urban | 1.14 | 1.165 | 0.2 | 0.2 | 0.085 | 0.087 |
| ◻ Suburban | 1.16 | 2.31 | 0.2 | 0.38 | 0.087 | 0.173 |
| ◻ Rural | 1.71 | 11 | 0.28 | 1.9 | 0.13 | 0.86 |

**Separation Distance (km)**

**Fig. 5.13** The separation distance between WiMAX and FSS (36 MHz) for $\Delta f = 0$, 28 and 40 MHz with null synthesized method

Figure 5.12 depicts the minimum separation distances from 20 MHz WiMAX BS to 0.23 MHz FSS when a 20 dB shielding attenuation is used. It shows the CCI, 0 and 12 MHz guard band interference cases. The worst separation distance occurs in the CCI scenario for 5 m FSS receiver's antenna height in rural area environment. Separation distances listed in Fig. 5.12 are the only reasonable separations when a − 10 dB of $I/N$ protection ratio is considered. The additional isolation level of 72 dB is needed in the case of CCI scenario, and notably it is possible to reach this power reduction with null synthesized technique. When a typical channel of 36 MHz bandwidth is used to determine the minimum separation distance towards the WiMAX BS, it is found that the coexistence becomes even more possible according to the acceptable separation distance obtained in Fig. 5.13.

The figure above shows that the dense urban environment is the most appropriate area for coexistence. This is intuitive because the higher the clutter losses, the higher the attenuation and the interfering signal, from which a 40 dB attenuation type of shielding is highly recommended for the coexisted systems, especially in the case of co-channel systems. Furthermore, with 0 and 12 MHz guard bands, coexistence with a short separation distance is proved possible except for the case of 0.23 MHz FSS bandwidth (Fig. 5.14).

On the other hand, for the FSS receiver (12 MHz guard band, 1.8 m antenna height, 40 dB shielding attenuation and operating at dense urban area deployment), the introduction of fixed null synthesizer onto its neighbouring FSS BS imposes

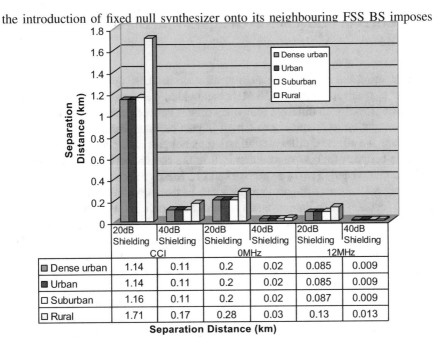

| | 20dB Shielding | 40dB Shielding | 20dB Shielding | 40dB Shielding | 20dB Shielding | 40dB Shielding |
| --- | --- | --- | --- | --- | --- | --- |
| | CCI | | 0MHz | | 12MHz | |
| ☐ Dense urban | 1.14 | 0.11 | 0.2 | 0.02 | 0.085 | 0.009 |
| ■ Urban | 1.14 | 0.11 | 0.2 | 0.02 | 0.085 | 0.009 |
| ☐ Suburban | 1.16 | 0.11 | 0.2 | 0.02 | 0.087 | 0.009 |
| ☐ Rural | 1.71 | 0.17 | 0.28 | 0.03 | 0.13 | 0.013 |

**Separation Distance (km)**

**Fig. 5.14** The separation distance between WiMAX and FSS (1.8 m height and 36 MHz) for $\Delta f = 0$, 28 and 40 MHz with null synthesized method under 20 and 40 dB shielding attenuation

minimum separation distances of 0.1 km and 0.009 km corresponding to the bandwidths of 0.23 MHz and 36 MHz, respectively. Interestingly, for a WiMAX BS and FSS receiver to coexist, a minimum separation distance that is less than 200 m is a prerequisite along with 1.8 m FSS receiver antenna (for 12 MHz guard band protection for 0.23 MHz FSS channel bandwidth). While for the 36 MHz FSS bandwidth, it can be noticed that successful coexistence can be achieved with 1.8 m FSS receiver antenna height for both 0 MHz and 12 MHz guard bands.

It can be noted that the antenna height has a considerable impact on the coexistence parameters together with deployment areas. Once again, the rural area environment exhibits the highest interference levels among all other operating environments. It can be clearly observed that in the rural area, the separation distance is nearly constant. However, increasing the clutter heights will definitely affect the clutter loss values.

At the same time, a fixed loss will be obtained when FSS antenna is higher than the clutter heights. Therefore, the rural areas may be considered as the poorest environment for coexistence investments. In contrast to rural areas, dense urban areas are the most feasible for coexistence because of its highest clutter loss values.

## 5.8   Effect of Mitigation Techniques with Different Guard Band

By comparing the results of separation distance in dense urban deployment with 1.8 m FSS receiver height and 36 MHz bandwidth, it is found that the coexistence can be improved with different frequency offsets between carriers. Therefore, proposed mitigation techniques have a high impact on interference. Figures 5.15 and 5.16 show the effect of different mitigation techniques on the minimum separation distance between IMT-Advanced and FSS in the co-channel interference scenario (0 MHz guard band, 5 MHz guard band and 12 MHz guard band).

Figure 5.15 shows that harmonization can be done even in the co-channel interference scenario when both mitigation techniques are used. However, it is considered that the effective isotropic power of the interferer is reduced in the direction of 1.8 m fixed-satellite earth station. On the FSS side, the shielding has attenuated the incoming interference by considering the clutter loss of dense urban area. This shows satisfactory results in terms of terrestrial signal attenuation to reduce the harmful interference.

The results presented in Figs. 5.16, 5.17 and 5.18 indicate that the mitigation techniques gave the same percentage of the distance reduction for both CCI and ACI scenarios. Additionally, the guard band plays an important role in the coexistence scenario. However, adjusting a sufficient guard band mitigation technique can yield a significant result in achieving the coexistence with different victim bandwidths and heights.

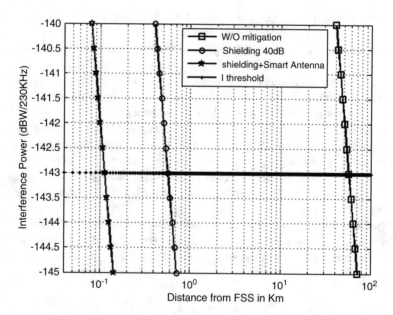

**Fig. 5.15** The effect of mitigation techniques on the minimum separation distance between IMT-Advanced and FSS in the co-channel interference scenario

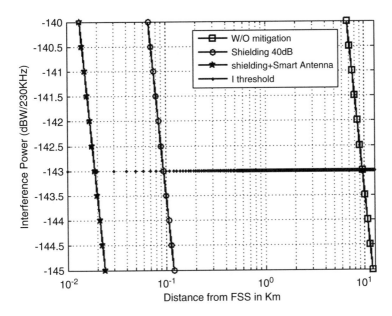

**Fig. 5.16** The effect of mitigation techniques on the minimum separation distance between IMT-Advanced and FSS using 0 MHz guard band

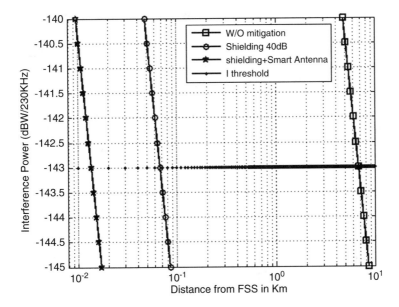

**Fig. 5.17** The effect of mitigation techniques on the minimum separation distance between IMT-Advanced and FSS using 5 MHz guard band

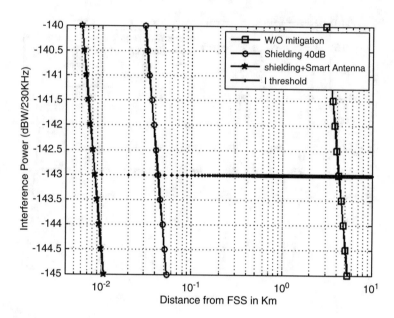

**Fig. 5.18** The effect of mitigation techniques on the minimum separation distance between IMT-Advanced and FSS using 12 MHz guard band

## 5.9 The Coexistence Possibilities of Using 40 dB Shielding and Fixed Null Within 0.2 km Separation Distance

According to the nature of WiMAX applications, the longest service interruption may not be more than 0.2 km. Consequently, in this book, a 0.2 km distance has been considered as a standard for feasibility in coexistence to suit the requirements of FSS receiver protection in this book.

Table 5.3 shows the separation distance required for coexistence in four different deployment areas. The grey block represents a possible coexistence, while the blank block indicates that deployment with 0.2 km separation is impossible.

The results indicate the possibility of IMT-Advanced (20 MHz) and FSS (0.23 MHz) ESs coexisting in the same geographical area. Additional isolation level of 72 dB is difficult to achieve even with the mitigation techniques. This case occurs when 5 m antenna height was used in the CCI scenario for less than 0.2 km horizontal separation.

It is found that within the 12 MHz guard band separation between the two services, coexistence is possible within the mitigation techniques for 36 MHz FSS bandwidth. In a case when a 36 MHz FSS bandwidth is used, the 0 MHz guard band separation between the two services of the coexistence is possible within the mitigation techniques except in the rural area, unless the FSS height was 1.8 m. In case of 0.23 MHz FSS bandwidth, coexistence with the 20 MHz WiMAX is possible with 12 MHz guard band separation, except for the rural area, unless the FSS height was 1.8 m.

**Table 5.3**  Summary of coexistence within 0.2 km

| 0.2 km standard | | Victim FSS antenna height (m) | | | | | | | |
|---|---|---|---|---|---|---|---|---|---|
| | | Dense urban | | Urban | | Sub-urban | | rural | |
| FSS Bandwidth | Interference scenarios | 1.8 | 5 | 1.8 | 5 | 1.8 | 5 | 1.8 | 5 |
| 0.23 MHz | Co-Channel | | | | | | | | |
| | 0 MHz GB | | | | | | | | |
| | 12 MH GB | X | X | X | X | X | X | X | |
| 36 MHz | Co-Channel | X | X | X | X | X | | X | |
| | 0 MHz GB | X | X | X | X | X | X | X | X |
| | 12 MH GB | X | X | X | X | X | X | X | X |

The fixed null synthesized method has been applied to reduce the separation distance in the CCI and ACI scenarios in a different shielding attenuation and different areas (rural, suburban, urban and dense urban area). By selecting a reasonable separation distance to represent the feasibility, it is important to evaluate the available chances for coexistence between two services.

## 5.10   Summary

The I-MUSIC algorithm is derived for the blind angle estimation, which avoids the high computational process of other subspace-base algorithms. It has been demonstrated that I-MUSIC can have much better performance for angle estimation in contrast to MUSIC, MI-MUSIC and MUSIC-Like algorithms. The algorithm can also work very well in other array manifolds.

The proposed method is based on a beam cancellation for frequency sharing between IMT-Advanced base stations (BS) and FSS in the 3400–4200 MHz frequency range. The method is significantly capable of steering a direct power towards the user by simulation. At the same time, the nulls remain fixed in the direction of FSS receiver. The fixed null synthesized method has been applied to reduce the separation distance in the co-channel and adjacent channel interference scenarios. It is applied with a different shielding attenuation and different areas (rural, suburban, urban and dense urban area).

A proven approach to solve the coexistence problems of WiMAX BS with FSS receiver is introduced in this book. The proposed approach is a combination of two well-known algorithms, namely, I-MUSIC and LMS. It is blended together in order to generate beams in the direction of WiMAX users while nulling those towards FSS receivers. The method is proven efficient in reducing the separation distance in both co-channel and adjacent channel interference scenarios with various shielding attenuation and at different deployment environments.

Among the tested scenarios, the one in which the FSS antenna height is 1.8 m proved the best configuration to be deployed. Remarkably, the null synthesizer alongside the 40 dB shielding is introduced to reduce the separation distance by 99.645% compared to the 0 dB shielding with no null synthesizer situation.

A novel method has been introduced in order to predict the systems coexistences caused by interference between services in several deployment areas. Several methods of investigating the signal, co-channel, zero-guard band and adjacent channel, with 0 dB, 20 dB and 40 dB shielding attenuation, have been proposed to investigate the phenomenon (in the frequency, power and space domains to obtain the correlation between the minimum separated range of base station antennas and the frequency separation). Different channel bandwidths have been considered in the method for the victim FSS to further validate the results in terms of various applications of satellite receivers.

# References

1. K. Huarng, C. Yeh, A unitary transformation method for angle of arrival estimation. IEEE Trans. Acoust. Speech, Signal Process. **39**(4), 975–977 (1991)
2. M. Pesavento, A. Gershman, M. Haardt, On unitary root-MUSIC with a real-valued eigende-composition: a theoretical and experimental performance study. IEEE Trans. Signal Process. **48**(5), 1306–1314 (2000)
3. G.V. Tsoulos, G.E. Athanasiadou, *Adaptive Antenna Arrays for Mobile Communications: Performance/System Considerations and Challenges* (COMCON '99, Athens, 1999)
4. A.O. Boukalov, S.G. Haggman, System aspects of smart-antenna technology in cellular wireless communications—An overview. IEEE Trans. Microwave Theory Tech. **48**(6), 919–928 (2000)
5. G.V. Tsoulos, Smart antennas for mobile communication systems: benefits and challenges. Electron. Commun. Eng. J **11**, 84–94 (1999)
6. 3GPP2/TSG-C.R1002, *1xEV-DV Evaluation Methodology (V12.1)*, NOKIA, (2003)
7. H.-S. Jo et al., The coexistence of OFDM-based systems beyond 3G with fixed service microwave systems. J. Commun. Netw. **8**(2), 187–193 (2006)
8. R. Schmidt, Multiple emitter location and signal parameter estimation. IEEE Trans. Antennas Propag. **34**(3), 276–280 (1986)
9. X. Zhang, G. Feng, D. Xu, Blind direction of angle and time delay estimation algorithm for uniform linear array employing multi-invariance music. Progr. Electromagn. Res. Lett. **13**, 11–20 (2010)
10. T.B. Lavate, V.K. Kokate, A.M. Sapkal, Performance analysis of MUSIC and ESPRIT DOA estimation algorithms for adaptive array smart antenna in mobile communication. Int. J. Comput. Netw. (IJCN) **2**(3), 152–158 (2010)
11. Recommendation ITU-R P.452-12, *Prediction Procedure for the Evaluation of Microwave Interference Between Stations on the Surface of the Earth at Frequencies Above About 0.7 GHz*, Geneva, Switzerland, May 2007

# Chapter 6
# Conclusions and Future Work

## 6.1 Conclusions

The results of this study provide guidelines and construction tools for coexistence possibilities between terrestrial services (such as IMT-Advanced) and satellite services (such as VSAT receiver) in the 3400–4200 MHz frequency range. The main findings of this book are summarized below, and they may be used either as a starting point for future researches or as an outline for system designers.

Significant experiment studies and simulations have been performed. The results of the studies have been used to propose an improved mitigation technique for reducing the interference between FSS receiver and IMT-Advanced system. Various factors such as separation distance, frequency separation, transmitted power, transmitter and receiver characteristics, propagation model and intersystem interference have been identified and clearly demonstrated. Additionally, shielding technique, smart antenna elements and MUSIC algorithm background and concept have been highlighted and discussed in details as possible mitigation techniques. It is concluded that an improved shielding technique is needed to achieve the minimum separation distance. Furthermore, the study has demonstrated the applicability of a practical mitigation technique which could further increase the possibility of sharing between these systems using guard band inserted between the two services.

It is also concluded that the antenna gain is an important factor to achieve the feasible coexistence. Therefore, one possible method is to apply smart antennas on IMT systems in order to null the EIRP in the direction of the interference with the FSS earth station.

The methodological approach of this study is based on the protection ratio of FSS receiver. It is found that the receiver's adjacent channel selectivity reduction is unnecessary in the CCI scenario when both of the services have the same bandwidth. However, if the interferer's bandwidth is larger than that of the victim's, then another factor should be added to account for mask discrimination correction.

© Springer International Publishing AG 2018
L.F. Abdulrazak, *Coexistence of IMT-Advanced Systems for Spectrum Sharing with FSS Receivers in C-Band and Extended C-Band*,
https://doi.org/10.1007/978-3-319-70588-0_6

It is observed that the separation distance decreases with an increasing channel bandwidth. However, higher bandwidth implies higher noise in the receiver and consequently a higher noise floor level. As a result, the interfering signal strength (dBm) becomes stronger as the distance is nearer. Therefore, the interference becomes more visible when the interferer bandwidth is greater than that of victim. On the other hand, less interference is encountered when the victim receiver bandwidth is wider than that of interfering transmitter. This is due to the adjacent channel leakage ratio, which accounts for the fact that wider bandwidth of interferer results in lower spectral emissions, especially in the adjacent bands. Moreover, the proposed approach presents a tractable and systematic workflow for the calculation of protection ratio. The proposed technique is also applicable to a wide range of frequencies and bandwidths by simply calculating the required threshold degradation, ACLR and ACS.

It has been found that if the actual terrain propagation conditions, such as the influence of artificial objects, are taken into consideration, the required separation distance will be reduced, and the interference area becomes spatially limited. This clearly justifies the increasing possibility of the sharing between FSS and IMT systems.

On the shielding mitigation technique, it was found that different materials have different levels of signal attenuation. The proposed shielding material (0.1 mm thickness zinc sheet) was a balanced choice, providing a high attenuation (about 20 dB) at a lower cost, compared to other metals. It was also found that it is possible to reduce the harmful interference up to 10% by increasing the shielding attenuation up to 20 dB. Consequently the separation distance can be minimized to 1% for 40 dB shielding attenuation.

The interference assessment from WiMAX to FSS has been verified in terms of guard band, antenna discrimination, shielding, different FSS bandwidth and diverse deployment areas in order to achieve minimum separation distance for each scenario. Simulations of the coexistence effect, up to unlimited frequency band separation, can be achieved by using the adjacent channel interference model.

It was also found that adjusting the interfering EIRP can reduce the transmitter power and antenna gain towards the FSS receiver antenna. This could be translated to interference reduction. This is important in order to create the fixed null extraction technique since the victim receiver gain depends on the direction of interference source. Consequently, reducing the receiver gain will help mitigate the interference. Therefore, increasing the antenna off-axis angle or down tilt, and using site engineering to position the antenna, will null the radiation pattern pointing towards the interference source. This is not practically achievable with FSS because the FSS receiver normally points to a fixed position of GSO and the off-axis angle is not variable for one position unless multi GSOs are available. Therefore, the transmitter gain is the only parameter that can be changed using site engineering to mitigate the harmful interference. It can be concluded that the dense urban and urban areas are the best areas for coexistence and intersystem interference coordination. However, the existence of buildings, trees, high number of users and other objects can reduce the size of IMT-Advanced base station.

From the smart antenna point of view, a novel method was used to increase the resolution of DOA algorithm using the I-MUSIC. It is then inserted with the

information of FSS ES in the beamforming of LMS algorithm. This full mechanism is capable of steering the adaptive beam towards the IMT-Advanced user
while keeping the fixed null in the direction of FSS ES. It is therefore concluded
that interference in the co-channel between two services could be achieved for a
36 MHz FSS bandwidth by combining the 40 dB shielding attenuation and null
synthesized method, at 1.8 m antenna height. However, a very small FSS bandwidth such as 0.23 MHz can only coexist with the 12 MHz WiMAX when FSS
receiver height is 1.8 m in the dense urban areas.

These proposed techniques have shown all the evidences that coexistence within
a feasible separation distance is possible. Also the technique can reduce down the
separation distance between two systems to 0.335% of the original separation distance with 40 dB shielding. The proposed mitigation technique is much related to
the immerging technology in wireless communication. However, the technique can
further be extended to support future studies which are important at both national
and international levels. It can also be used to support the frequency administrators
and regulatory bodies in order to achieve best coexistence between FSS receiver and
IMT-Advanced.

Finally, by using the resultant techniques mentioned in this book, frequency
sharing between FSS and IMT-Advanced systems in 3400–4200 MHz band within
optimum separation distance can be designed and implemented under certain conditions. Therefore, the shielding and fixed nulls have been used to reduce the harmful
power interference from the IMT-Advanced base station to the FSS receiver and
also to achieve the minimum separation distance. This will lead to an efficient use
of limited frequency spectrum resources.

## 6.2   Future Work

This book has posed some research questions which need further investigations. It
is recommended that further studies be undertaken in the following areas:

- Future trials are needed in order to determine a better technique that can be
  developed using different strategies and materials to protect the FSS receiver.
- Further researches in the field of the coexistence role using different models as
  mentioned in Chap. 2 would be of great help in different systems. Further inves-
  tigation and experimentation with different interferer bandwidths are strongly
  recommended. In particular, it will be of great interest to conduct investigations
  and comparison of different models. This will enable the radio engineers to select
  the most suitable propagation model for wider range of wireless applications.
- The idea of considering the FSS as a receiver only can be changed by adding a
  subscription card in the FSS-ES to send an IMT-Advanced beacon signal. By
  doing so, IMT-BS will determine the DOE and avoid the CCI and ACI by
  developing a mitigation technique based on cognitive radio technology to mitigate the interference. This can be achieved either by frequency carrier shifting or
  beamforming.

# Appendix A: List of Author's Related Publications

**Chapters in Books**

1. **Lway Faisal Abdulrazak.**, Kusay Faisal Al-Tabatabaie., and Tharek Abd. Rahman. *Utilize 3300–3400 MHz Band for Fixed Wireless Access.* In *"Advanced Technologies"* book. INTECH publications. ISBN 978–953–307–017–9. pp.291–301.
2. Zaid A. Shamsan, **Lway F.** and Tharek Abd Rahman, *Co-sited and Non Co-sited Coexistence Analysis between IMT-Advanced and FWA Systems in Adjacent Frequency Band.* In *New Aspects Of Telecommunications And Informatics*, Published by WSEAS Press 2008, pp. 44–48. ISBN: 978–960–6766–64–0, ISSN: 1790–5117.
3. Zaid A. Shamsan, **Lway Faisal** and Tharek Abd. Rahman. *Transmit Spectrum Mask for Coexistence between Future WiMAX and Existing FWA Systems.* In *Recent Advances In Systems Engineering And Applied Mathematics*, Published by WSEAS Press 2008, pp. 76–83. ISBN: 978–960–6766–91–6, ISSN 1790–2769.

**International Journals**

1. **Lway Faisal Abdulrazak**, Zaid A. Shamsan and Tharek Abd. Rahman. Potential Penalty Distance between FSS Receiver and FWA for Malaysia. *International Journal Publication in WSEAS Transactions on COMMUNICATIONS*, ISSN: 1109–2742, Issue 6, Volume 7, June 2008, pp. 637–646.
2. **Lway Faisal Abdulrazak** and Tharek Abd. Rahman. Introduce the FWA in the band 3300–3400 MHz. *Proceedings of World Academy of Science, Journal of Engineering and Technology (PWASET)*, Volume 46–32, December 2008 ISSN 2070–3740. pp. 176–179.
3. Zaid A. Shamasn, **Lway Faisal** and Tharek Abd. Rahman. On Coexistence and Spectrum Sharing between IMT-Advanced and Existing Fixed Systems.

© Springer International Publishing AG 2018
L.F. Abdulrazak, *Coexistence of IMT-Advanced Systems for Spectrum Sharing with FSS Receivers in C-Band and Extended C-Band*,
https://doi.org/10.1007/978-3-319-70588-0

*International Journal Publication in WSEAS TRANSACTIONS on COMMUNICATIONS*, Issue 5, Volume 7, May 2008, pp. 505–515.

4. Zaid A. Shamasn, **Lway Faisal**, S. K. S.Yusof, Tharek A. Rahman. Spectrum Emission Mask for Coexistence between Future WiMAX and Existing Fixed Wireless Access Systems. *International Journal Publication in WSEAS Transactions on COMMUNICATIONS*, Vol. 7, Issue 6, June 2008, pp. 627–636.

5. **Lway Faisal Abdulrazak**, Arshed A. O. Tractable Technique to Evaluate the Terrestrial to Satellite Interference in the C-Band Range. *International Journal of Theoretical and Applied Information Technology*, Vol. 65, No.3, pp. 762–769. 2014.

6. **Lway Faisal Abdulrazak** and Arshed A. O. Interference Mitigation Technique through Shielding and Antenna Discrimination. *International Journal of Multimedia and Ubiquitous Engineering.* Vol. 10, No. 3 (2015), pp. 343–352.

7. **Lway Faisal Abdulrazak,** Kusay F. Al-Tabatabaie. Broad-Spectrum Model for Sharing Analysis between IMT-Advanced Systems and FSS Receiver. *Journal of Electronics and Communication Engineering (IOSR-JECE),*Volume 12, Issue 1, Ver. III (Jan.-Feb. 2017), pp. 52–56.

8. **Lway Faisal Abdulrazak,** Kusay F. Al-Tabatabaie. Preliminary design of iraqi spectrum management software (ISMS), *International Journal of Advanced Research.* Vol.5(issue2), pp. 2560–2568.

**Conference Papers**

1. **Lway Faisal Abdulrazak**, Tharek Abd Rahman. Review Ongoing Research of Several Countries on the Interference between FSS and BWA. *International Conference on Communication Systems and Applications (ICCSA'08)*, 2008 Hong Kong China.

2. **Lway Faisal Abdulrazak** and Tharek Abd. Rahman. Novel Computation of Expecting Interference between FSS and IMT-Advanced for Malaysia. *2008 IEEE International RF and Microwave Conference (RFM08)*, Malaysia 2–4th December 2008.

3. **Lway Faisal Abdulrazak**, S.K.Abdul Rahim, and Tharek Abd. Rahman. New Algorithm to Improve the Coexistence between IMT-Advanced Mobile Users and Fixed Satellite Service. *Proceedings of 2009 International IACSIT Conference on Machine Learning and Computing (IACSIT ICMLC 2009)* Perth, Australia, July 10–12, 2009.

4. **Lway Faisal Abdulrazak**, Zaid A. Hamid, Zaid A. Shamsan, Razali Bin Nagah and Tharek Abd. Rahman. The Co-Existence of IMT-Advanced And Fixed Satellite Service Networks In The 3400–3600 MHz. *Proceeding of MCMC colloquium 2008.* 18-19December 2008. ISBN: 978–983–42,563–2–6. pp. 77–82.

5. **Lway F. Abdulrazak**, Zaid A. Shamsan and Tharek Abd. Rahman. Potentiality of Interference Correction between FSS and FWA for Malaysia. *Selected Paper from the World Scientific and Engineering Academy and Society Conferences* in Istanbul, Turkey, May 2008, pp. 84–89.

6. **Lway Faisal Abdulrazak**, Tharek Abd. Rahman, and S.K.Abdul Rahim. IMT-Advanced and FSS Interference Area Ratio Methodology. *Proceeding of the 8th international conference on circuits, systems, electronics, control & signal processing (CSECS'09).* December 14–16,2009. ISBN: 978–960–474–139–7, ISSN:1790–5117.

7. Zaid A. Shamsan, **Lway Faisal**, and Tharek Abd Rahman. Co-sited and Non Co-sited Coexistence Analysis between IMT-Advanced and FWA Systems in Adjacent Frequency band. *in Proceedings of the International Conference on Telecommunications and Informatics (TELE-INFO '08).* pp. 44–48, Istanbul, Turkey, May 27–30, 2008.

8. Zaid A. Shamsan, **Lway Faisal** and Tharek Abd. Rahman. Transmit Spectrum Mask for Coexistence between Future WiMAX and Existing FWA Systems. *Selected Paper from the World Scientific and Engineering Academy and Society Conferences* in Istanbul, Turkey, May 2008, pp. 76–83.

9. Zaid A. Shamsan, **Lway F. Abdulrazak**, Tharek Abd. Rahman. On the Impact of Channel Bandwidths and Deployment Areas Clutter Loss on Spectrum Sharing of Next Wireless Systems. *International Conference on Electronic Design (ICED08)*, Malaysia 1–3 December 2008.

10. Shamsan Zaid A., **Faisal L.** and Rahman Tharek A. (2008). Compatibility and Coexistence between IMT-Advanced and Fixed Systems, *in Proceedings of the 4th International Conference and Information Technology and Multimedia at UNITEN (ICIMU' 2008)*, 17–19 November, Malaysia.

11. Shamsan Zaid A., **Abdulrazak Lway F.,** and Rahman Tharek A. (2008). Co-channel and Adjacent Channel Interference Evaluation for IMT-Advanced Coexistence with Existing Fixed System, *in Proceedings of IEEE International RF and Microwave Conference (RFM 2008),* pp. 65–69, 2–4 December, Malaysia.

# Appendix B: MEASAT-3 Satellite for C-Band

## Appendix B.1: MEASAT-3 Satellite Network

Binariang Satellite Systems Sdn Bhd, operator of MEASAT-1 and MEASAT-2 satellite networks, has launched their third satellite network in the first quarter of 2007. This satellite network is co-located with MEASAT-1 at 91.5° east and providing additional capacity for the current satellite; and was the replacement satellite when MEASAT-1 was took out of service; as well as to provide capacity for restoration of the existing satellite in orbit. Frequency bands filed for MEASAT-3 satellite network in the C-band and Ku-band are shown in Table B.1.

MEASAT-3 is operating on Ku-band, C-band and also extended C-band. The frequency configuration is shown in Fig. B.1 and Table B.2. The area covered by C-band as shown in Figs. B.2, B.3 and B.4 are: Southeast Asia beam, global beam and coverage area that use Ku-band.

**Table B.1** MEASAT-3 frequency bands filed

| MEASAT networks | Uplink frequency (MHz) | Downlink frequency (MHz) | Type of service |
|---|---|---|---|
| MEASAT-3 | 5925–6725 | 3400–4200 | Fixed satellite |
| | 7900–8400 | 7250–7750 | Fixed satellite |
| | 13,750–14,500 | 10,950–11,200<br>11,450–11,700<br>12,200–12,750 | Fixed satellite |

© Springer International Publishing AG 2018
L.F. Abdulrazak, *Coexistence of IMT-Advanced Systems for Spectrum Sharing with FSS Receivers in C-Band and Extended C-Band*,
https://doi.org/10.1007/978-3-319-70588-0

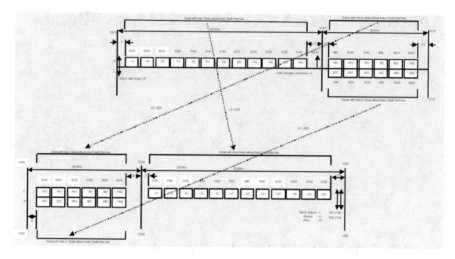

**Fig. B.1** MEASAT-3 C-band channelling

**Fig. B.2** MEASAT-3 C-band EIRP (dBW) Southeast Asia footprint

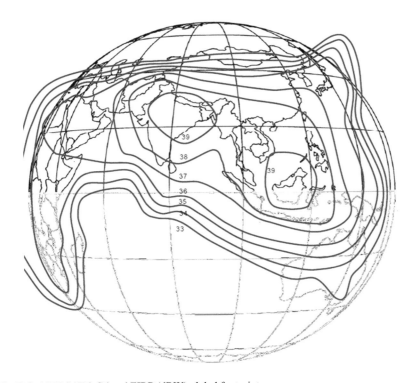

**Fig. B.3**  MEASAT-3 C-band EIRP (dBW) global footprint

**Table B.2**  MEASAT-3 C-band frequency configuration

| MEASAT networks polarization | Uplink frequency (MHz) | Downlink frequency (MHz) | Area of service coverage |
|---|---|---|---|
| Vertical | 6425–6725 | 3400–4200 | Southeast Asia including Malaysia |
| Horizontal | 5925–6725 | 3400–3700 | Southeast Asia including Malaysia |
| Vertical | 6425–6665 | 3400–4200 | C-band global without India |
| Horizontal | 5925–6665 | 3400–3700 | C-band global without India |
| Vertical | 6425–6665 | 3400–4200 | C-band global without India |
| Horizontal | 5925–6665 | 3400–3700 | C-band global without India |
| Vertical | 14,250–14,500 | 10,950–12,750 | Ku-band over Malaysia |
| Horizontal | 13,750–14,500 | 10,950–11,700 | Ku-band over Malaysia |

**Fig. B.4** MEASAT-3 Ku-band EIRP (dBW) footprints

# Appendix C: Mathematical Equations

## Appendix C.1: The Eigen-Decomposition: Eigenvalues and Eigenvectors

Eigenvectors and eigenvalues are numbers and vectors associated to square matrices, and together they provide the eigen-decomposition of a matrix which analyses the structure of this matrix. Even though the eigen-decomposition does not exist for all square matrices, it has a particularly simple expression for a class of matrices often used in multivariate analysis such as correlation, covariance or cross-product matrices. The eigen-decomposition of this type of matrices is important in statistics because it is used to find the maximum (or minimum) of functions involving these matrices. For example, principal component analysis is obtained from the eigen-decomposition of a covariance matrix and gives the least square estimate of the original data matrix. Eigenvectors and eigenvalues are also referred to as characteristic vectors and latent roots or characteristic equation (in German, "eigen" means "specific of" or "characteristic of"). The set of eigenvalues of a matrix is also called its spectrum.

There are several ways to define eigenvectors and eigenvalues; the most common approach defines an eigenvector of the matrix $A$ as a vector $u$ that satisfies the following equation:

$$Au = \lambda u$$

When rewritten, the equation becomes:

$$(A - \lambda I)u = 0$$

where $\lambda$ is a scalar called the eigenvalue associated to the eigenvector. In a similar manner, we can also say that a vector $u$ is an eigenvector of a matrix $A$ if the length of the vector (but not its direction) is changed when it is multiplied by $A$.

L.F. Abdulrazak, *Coexistence of IMT-Advanced Systems for Spectrum Sharing with FSS Receivers in C-Band and Extended C-Band*,
https://doi.org/10.1007/978-3-319-70588-0

For example, the matrix $A = \begin{vmatrix} 2 & 3 \\ 2 & 1 \end{vmatrix}$ has the eigenvectors $u_1 = \begin{vmatrix} 3 \\ 2 \end{vmatrix}$, with eigenvalue $\lambda_1 = 4$, and $u_2 = \begin{vmatrix} -1 \\ 1 \end{vmatrix}$ with eigenvalue $\lambda_2 = -1$. We can verify (as illustrated in Fig. C.1) that only the length of $u_1$ and $u_2$ is changed when one of these two vectors is multiplied by the matrix $A$:

$$\begin{vmatrix} 2 & 3 \\ 2 & 1 \end{vmatrix}\begin{vmatrix} 3 \\ 2 \end{vmatrix} = 4\begin{vmatrix} 3 \\ 2 \end{vmatrix} = \begin{vmatrix} 12 \\ 8 \end{vmatrix}$$

and

$$\begin{vmatrix} 2 & 3 \\ 2 & 1 \end{vmatrix}\begin{vmatrix} -1 \\ 1 \end{vmatrix} = -1\begin{vmatrix} -1 \\ 1 \end{vmatrix} = \begin{vmatrix} 1 \\ -1 \end{vmatrix}$$

For most applications, we normalize the eigenvectors (i.e. transform them such that their length is equal to one):

$$u^T u = 1$$

For the previous example, we obtain:

$$u_1 = \begin{vmatrix} 0.8331 \\ 0.5547 \end{vmatrix}$$

We can check that:

$$\begin{vmatrix} 2 & 3 \\ 2 & 1 \end{vmatrix}\begin{vmatrix} 0.8331 \\ 0.5547 \end{vmatrix} = \begin{vmatrix} 3.3284 \\ 2.2188 \end{vmatrix} = 4\begin{vmatrix} 0.8331 \\ 0.5547 \end{vmatrix}$$

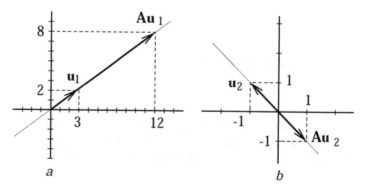

**Fig. C.1** Two eigenvectors of a matrix

and

$$\begin{vmatrix} 2 & 3 \\ 2 & 1 \end{vmatrix}\begin{vmatrix} -0.7071 \\ 0.7071 \end{vmatrix} = \begin{vmatrix} 0.7071 \\ -0.7071 \end{vmatrix} = -1\begin{vmatrix} -0.7071 \\ 0.7071 \end{vmatrix}$$

Traditionally, we put together the set of eigenvectors of $A$ in a matrix denoted $U$. Each column of $U$ is an eigenvector of $A$. The eigenvalues are stored in a diagonal matrix (denoted $\Lambda$), where the diagonal elements give the eigenvalues (and all the other values are zeros). We can rewrite the first equation as:

$$AU = \Lambda U$$

or also as:

$$A = U\Lambda U^{-1}$$

For the previous example, we obtain:

$$A = U\Lambda U^{-1}$$

$$= \begin{vmatrix} 3 & -1 \\ 2 & 1 \end{vmatrix}\begin{vmatrix} 4 & 0 \\ 0 & -1 \end{vmatrix}\begin{vmatrix} 2 & 2 \\ -4 & 6 \end{vmatrix} = \begin{vmatrix} 2 & 3 \\ 2 & 1 \end{vmatrix}$$

It is important to note that not all matrices have eigenvalues. For example, the matrix $\begin{vmatrix} 0 & 1 \\ 0 & 0 \end{vmatrix}$ does not have eigenvalues. Even when a matrix has eigenvalues and eigenvectors, the computation of the eigenvectors and eigenvalues of a matrix requires a large number of computations and is therefore better performed by computers.

## Appendix C.2: Conjugate Matrix

A conjugate matrix is a matrix $A^*$ obtained from a given matrix $A$ by taking the complex conjugate of each element of $A$:

$$\overline{\left(a_{ij}\right)} = \left(\overline{a_{ij}}\right)$$

The notation $A^*$ is sometimes also used, which can lead to confusion since this symbol is also used to denote the conjugate transpose.

## Appendix C.3: Conjugate Transpose

The conjugate transpose of a $m \times n$ matrix $A$ is the $n \times m$ matrix defined by:

$$A^H \equiv \overline{A}^T$$

where $A^T$ denotes the transpose of matrix $A$ and $\overline{A}$ denotes the conjugate matrix. In all common space, the conjugate and transpose operations commute, so:

$$A^H \equiv \overline{A}^T = \overline{A^T}$$

## Appendix C.4: Frobenius Norm

The Frobenius norm, sometimes also called the Euclidean norm (which may cause confusion with the vector $L^2$ norm also sometimes known as the Euclidean norm), is a matrix norm of a $m \times n$ matrix $A$ defined as the square root of the sum of the absolute squares of its elements:

$$A_F \equiv \sqrt{\sum_{i=1}^{m} \sum_{j=1}^{n} |a_{ij}|^2}$$

## Appendix C.5: Root Mean Square Error (RMSE)

The RMSE is the square root of the variance of the residuals. It indicates the absolute fit of the model to the data – how close the observed data points are to the model's predicted values. Whereas R-squared is a relative measure of fit, RMSE is an absolute measure of fit. As the square root of a variance, RMSE can be interpreted as the standard deviation of the unexplained variance and has the useful property of being in the same units as the response variable. Lower values of RMSE indicate better fit. RMSE is a good measure of how accurately the model predicts the response and is the most important criterion for fit if the main purpose of the model is prediction:

$$RMSE = \sqrt{E[\overline{x} - x]^2}$$

where $E$ is the expectation, $\overline{x}$ is the mean and $x$ is the variable element.

## Appendix C.6: The Expectation

It is an arithmetic average, just one calculated from probabilities instead of being calculated from samples. So, for example, if $P(k)$ is the probability that we find $K A_1$ alleles in our sample, the *expected number* of $A_1$ alleles in our sample is just:

$$E(k) = \sum kP = np$$

## Appendix C.7: The Standard Deviation $\sigma^2$

It is the square root of the variance, where the variance is the average of the squared differences from the mean. So the variance is same like the expectation, and the standard deviation is equal to the RMSE.

## Appendix C.8: The Autocorrelation

Autocorrelation is the cross-correlation of a signal with itself. Informally, it is the similarity between observations as a function of the time separation between them. It is a mathematical tool for finding repeating patterns, such as the presence of a periodic signal which has been buried under noise, or identifying the missing fundamental frequency in a signal implied by its harmonic frequencies. It is often used in signal processing for analysing functions or series of values, such as time-domain signals:

$$R(s,t) = \frac{E[(x_t - \mu_t)(x_s - \mu_s)]}{\sigma_t \sigma_s}$$

# Appendix D: Null Synthesized Algorithm for Minimum Separation Distance

## Appendix D.1: Interference Assessment Methodology Code Without Mitigation Technique

The following code is for rural area with a 180 dish diameter 0 shielding attenuation and 36 MHz channel bandwidth. In order to change the environment or dish parameters, the parameters in the code should be adjusted.

```
% IMT-Advanced parameters
f1=4.040; % frequencey carrier in GHz
Gt=18; % antenna gain before mitigation technique in dBm for
IMT-Advanced
Pt=43; %transmitted power of interferer IMT-Advanced in dBm
Gr=-10; %gain of receiver(victim)in dBi
I1=-143;    %Interference threshold in dBw/230kHz worst case
%%%%%%%%%%%%%%%%%%%%%%%%%%%%%%%%%%%%%%%%%%%%%%%%%%%%%%%%%%%%%%%%%%%
%%%
%isolation from site shielding
r1=0;  % in dB no shielding
r2=20;
r3=40;
%%%%%%%%%%%%%%%%%%%%%%%%%%%%%%%%%%%%%%%%%%%%%%%%%%%%%%%%%%%%%%%%%%%
%%%
% clutter loss parameters
%%%%%%%%%%%%%%%%%%%%%%%%%%%%%%%%%%%%%%%%%%%%%%%%%%%%%%%%%%%%%%%%%%%
%%%
dk=0.02;
ha=25;
h=1.8;
BwI=20;%bandwidth of WiMAX (interferer)
BWV=36;%bandwidth of victim FSS receiver
```

© Springer International Publishing AG 2018
L.F. Abdulrazak, *Coexistence of IMT-Advanced Systems for Spectrum Sharing with FSS Receivers in C-Band and Extended C-Band*,
https://doi.org/10.1007/978-3-319-70588-0

```matlab
close all;
%%%%%%%%%%%%%%%%%%%%%%%%%%%%%%%%%%%%%%%%%%%%%%%%%%%%%%%%%%%%%%%%%%%%%%%%%%%%
%%%
% Propagation Model base on ITU-R P.452
%%%%%%%%%%%%%%%%%%%%%%%%%%%%%%%%%%%%%%%%%%%%%%%%%%%%%%%%%%%%%%%%%%%%%%%%%%%%
%%%
%sepration distance befor the mitigation technique for different
shielding
%effects for Co-channel interference
I1=[-145:0.5:-140];
%correction band calculation
if (BwI>=BWV)
    corr_band=-10*log10(BwI/BWV);
  else
    corr_band=0;
end
Ah=10.25*exp(-dk)*(1-tanh(6*((h/ha)-0.625)))-0.33;
%Co-Channel Interference Scenario
maskAtt=-34.5;
d1=(10.^(((-I1+Pt+maskAtt+corr_band+Gt+Gr-r1-Ah-92.5-
20*log10(f1))/20)));
i4=-143;
d5=[0:60];
%Zero_GB Interference Scenario
maskAtt=-50;
d2=(10.^(((-I1+Pt+maskAtt+corr_band+Gt+Gr-r1-Ah-92.5-
20*log10(f1))/20)));
%5MHz GB Interference Scenario
maskAtt=-53;
d3=(10.^(((-I1+Pt+maskAtt+corr_band+Gt+Gr-r1-Ah-92.5-
20*log10(f1))/20)));
%12MHz GB Interference Scenario
maskAtt=-57;
d4=(10.^(((-I1+Pt+maskAtt+corr_band+Gt+Gr-r1-Ah-92.5-
20*log10(f1))/20)));
 p2 = figure(2);
plot(d1,I1,'ks-','LineWidth',2);
hold on
plot(d2,I1,'ko-','LineWidth',2);
plot(d3,I1,'k--','LineWidth',2);
plot(d4,I1,'kp-','LineWidth',2);
 plot(d5,i4,'k.-','LineWidth',2);
grid on
title('Separation Distance for 1.8m FSS Receiving Antenna and
WiMAX Transmitter');
```

```
xlabel('Distance   from   FSS(Km)');   ylabel('Interference   Power
[dBW/230KHz)');
legend('Co-Channel Interference Scenario', 'Zero-GB Interference
Scenario', '5MHz GB Interference Scenario', '12MHz GB Interference
Scenario', 'I threshold');
```

## Appendix D.2: Interference Assessment Methodology Code After the Mitigation Technique

The following code is for rural area with a 180 dish diameter 20 shielding attenuation using null technique and 36 MHz channel bandwidth. In order to change the environment or dish parameters, the parameters in the code should be adjusted.

```
% IMT-Advanced parameters
f1=4.040; % frequencey carrier in GHz
Gant=4;
Gbf=0;
Gt=Gbf+Gant; % antenna gain before mitigation technique in dBm for
IMT-Advanced
Pt=43;  %transmitted power of interferer IMT-Advanced in dBm
Gr=-10; %gain of receiver(victim)in dBi
I1=-143;    %Interference threshold in dBw/230kHz worst case
%%%%%%%%%%%%%%%%%%%%%%%%%%%%%%%%%%%%%%%%%%%%%%%%%%%%%%%%%%%%%%%%%%%%%%
%%%
%isolation from site shielding
r1=20;
r3=40;
%%%%%%%%%%%%%%%%%%%%%%%%%%%%%%%%%%%%%%%%%%%%%%%%%%%%%%%%%%%%%%%%%%%%%%
%%%
% clutter loss parameters
%%%%%%%%%%%%%%%%%%%%%%%%%%%%%%%%%%%%%%%%%%%%%%%%%%%%%%%%%%%%%%%%%%%%%%
%%%
dk=0.02;
ha=25;
h=1.8;
BwI=20;%bandwidth of WiMAX (interferer)
BWV=36;%bandwidth of victim FSS receiver
close all;
%%%%%%%%%%%%%%%%%%%%%%%%%%%%%%%%%%%%%%%%%%%%%%%%%%%%%%%%%%%%%%%%%%%%%%
%%%
% Propagation Model base on ITU-R P.452
%%%%%%%%%%%%%%%%%%%%%%%%%%%%%%%%%%%%%%%%%%%%%%%%%%%%%%%%%%%%%%%%%%%%%%
%%%
```

```
%sepration distance befor the mitigation technique for different
shielding
%effects for Co-channel interference
I1=[-145:0.5:-140];
%correction band calculation
if (BwI>=BWV)
    corr_band=-10*log10(BwI/BWV);
  else
    corr_band=0;
end
Ah=10.25*exp(-dk)*(1-tanh(6*((h/ha)-0.625)))-0.33;
%Co-Channel Interference Scenario
maskAtt=-34.5;
d1=(10.^(((-I1+Pt+maskAtt+corr_band+Gt+Gr-r1-Ah-92.5-
20*log10(f1))/20)));
i4=-143;
d5=[0:2];
%Zero_GB Interference Scenario
maskAtt=-50;
d2=(10.^(((-I1+Pt+maskAtt+corr_band+Gt+Gr-r1-Ah-92.5-
20*log10(f1))/20)));
%5MHz GB Interference Scenario
maskAtt=-54;
d3=(10.^(((-I1+Pt+maskAtt+corr_band+Gt+Gr-r1-Ah-92.5-
20*log10(f1))/20)));
%12MHz GB Interference Scenario
maskAtt=-57;
d4=(10.^(((-I1+Pt+maskAtt+corr_band+Gt+Gr-r1-Ah-92.5-
20*log10(f1))/20)));
p2 = figure(2);
plot(d1,I1,'ks-','LineWidth',2);
hold on
plot(d2,I1,'ko-','LineWidth',2);
plot(d3,I1,'k--','LineWidth',2);
plot(d4,I1,'kp-','LineWidth',2);
plot(d5,i4,'k.-','LineWidth',2);
grid on
title('Separation Distance for 1.8m FSS Receiving Antenna and
WiMAX Transmitter');
xlabel('Distance   from   FSS(Km)');   ylabel('Interference   Power
[dBW/230KHz]');
legend('Co-Channel Interference Scenario', 'Zero-GB Interference
Scenario', '5MHz GB Interference Scenario', '12MHz GB Interference
Scenario', 'I threshold');
```

# Appendix D.3: I-MUSIC Spectrum Code for DOA Signal (Fig. D.1)

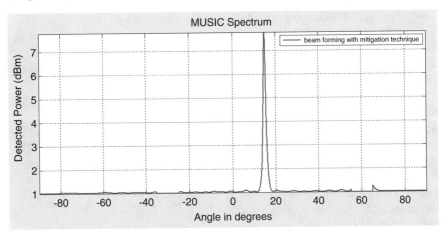

**Fig. D.1**  Signal Cancellation Using I-MUSIC Algorithm in the direction of FSS

## Appendix D.3.1: The Modification of I-MUSIC Code

```
clear;close all;clc;
num=8;%number of sourses
f=4e7;%working frequency
f1=3;%power of fist signal
f2=2;%power of second signal
f3=2;%power of fist signal
f4=3;%power of second signal
a1=75;%DOA of fist signal
a2=90;%DOA of second signal
a3=105;%DOA of fist signal
a4=120;%DOA of second signal
c=3e8;%light speed
lmda=c/f;
d=lmda/2;%spacing
thita=0:1:180;
SNR1=15;%signal to noise ratio
SNR2=15;%signal to noise ratio
SNR3=15;%signal to noise ratio
SNR4=15;%signal to noise ratio
fs=100;
Ts=[0:1/fs:1];
snapshot=length(Ts);%Pilot channel
```

```
s1=10^(SNR1/20)*sin(2*pi*f1*Ts);%1st signal wave
s2=10^(SNR2/20)*sin(2*pi*f2*Ts);%2nd signal wave
s3=10^(SNR3/20)*sin(2*pi*f1*Ts);%1st signal wave
s4=10^(SNR4/20)*sin(2*pi*f2*Ts);%2nd signal wave
athita1=exp(j*2*pi*d/lmda*cos(a1*pi/180)*[0:num-1]).';%steering
vector of first signal
athita2=exp(j*2*pi*d/lmda*cos(a2*pi/180)*[0:num-1]).';%steering
vector of secod signal
athita3=exp(j*2*pi*d/lmda*cos(a3*pi/180)*[0:num-1]).';%steering
vector of first signal
athita4=exp(j*2*pi*d/lmda*cos(a4*pi/180)*[0:num-1]).';%steering
vector of secod signal
x=athita1*s1+athita2*s2+sqrt(0.5)+athita3*s3+athita4*s4+sqrt(0.5)
*(randn(num,snapshot)+j*randn(num,snapshot));%x=s*a+n
R=x*x'/snapshot;
[V,D]=eig(R);
P=V(:,1:4)*V(:,1:4)';
e1=[1,zeros(1,num-1)].';
for i=1:length(thita)
    A=exp(j*2*pi*d/lmda*cos(thita(i)*pi/180)*[0:num-1]).';
    S(i)=1/(A'*P*A);
    S1(i)=1/(A'*P*e1);
end
figure(1)
% plot(thita,10*log10(abs(S)/(max(abs(S)))));
grid on;hold on;
plot(thita,10*log10(abs(S1)/(max(abs(S)))),'r--');
% title('I-MUSICspectrum')
xlabel('DOA Degree')
ylabel('Power Amplitude in dB')
% legend('commom-Music','Modified-Music')
legend('I-MUSICspectrum')
```

## Appendix D.3.2: The Modification of I-MUSIC Code with the Null Introduction

```
function edit10_Callback(hObject, eventdata, handles)
N= str2double(get(hObject,'String'));
handles.N = N;
guidata(hObject,handles)


function edit10_CreateFcn(hObject, eventdata, handles)
if ispc && isequal(get(hObject,'BackgroundColor'), get(0,'default
UicontrolBackgroundColor'))
```

```
     set(hObject,'BackgroundColor','white');
end
% --- Executes on button press in pushbutton1
function pushbutton1_Callback(hObject, eventdata, handles)

doas=[handles.th1 handles.th2 handles.th3]*pi/180;
P=[handles.p1 handles.p2 handles.p3];
r=length(doas);
% Steering vector matrix. Columns will contain the steering
vectors
% of the r signals
A=exp(-i*2*pi*handles.d*(0:handles.N-1)'*sin(doas));
% Signal and noise generation
sig=round(rand(r,handles.K))*2-1;% Generate random BPSK symbols
for each of the
% r signals
noise = sqrt ( handles . noise_
var/2)*(randn(handles.N,handles.K)+i*randn(handles.N,handles.K));
%Uncorrelated noise
X=A*diag(sqrt(P))*sig+noise;%Generate data matrix
R=X*X'/handles.K;%Spatial covariance matrix
[Q ,D]=eig(R);%Compute eigendecomposition of covariance matrix
[D,I]=sort(diag(D),1,'descend'); %Find r largest eigenvalues
Q=Q (:,I);%Sort the eigenvectors to put signal eigenvectors first
% Qs=Q (:,1:r);%Get the signal eigenvectors
Qn=Q(:,r+1:handles.N);%Get the noise eigenvectors
% MUSIC algorithm
% Define angles at which MUSIC "spectrum" will be computed
angles=(-90:0.1:90);
%Compute steering vectors corresponding values in angles
a1=exp(-i*2*pi*handles.d*(0:handles.N-1)'*sin(angles*pi/180));
for k=1:length(angles)
    %Compute I-MUSIC "spectrum"
e1=[1,zeros(1,D-1)].';

music_spectrum(k)=1/(a1(:,k)'*Qn*Qn'*e1(:,k));
end

exceptions=[-30 60];
for exception_time=1:length(exceptions)
    for ee=1:length(music_spectrum)
        if angles(ee)==exceptions(exception_time)
            music_spectrum(ee-5:ee+5)=0;
        end
```

```
        end
end

% for e=-90:0.1:90
%      angle=e*pi/180
% end
plot(angles,abs(music_spectrum),'LineWidth',2.0)
grid on;
legend('beam forming with mitigation technique');
title('MUSIC Spectrum'); ylabel('Normalized power') ; xlabel('Angle
in degrees'); axis tight;
```

## Appendix D.4: Beam Forming (Smart)

```
%*****************************************************************
*******
%   Beamforming_linear.m
%*****************************************************************
*******
%   It is a MATLAB function that simulates beamforming for linear
arrays.
%*****************************************************************
******
% Start timer
tic;
% Start recording
if (save_in_file=='y')
    out_file = sprintf('%s.txt',file_string);
    diary(out_file);
end
% User input
[N,d,sig,noise,type,nn,NN,AF_thresh,Mu,E_pattern] = linear_data_
entry;

% Parameters initialization
FIG         = 'figure(1)';              % Figure to record
SKIP_STEP   = 40;                       % Plot every SKIP_STEP
iterations
w                = zeros(N,1);                    % iteration
initialization
% Generate signals
[dd, X, fm] = linear_sig_gen(N,d,nn,NN,type,sig,noise,E_pattern);
```

```
for i = 1 : length(dd)
    w0 = w;
    [w, err(i)] = LMS(w,Mu,X(:,i),dd(i));
    mse(i) = sum(abs(err(i))^2);
    w_err(i) = norm(w0 - w);
    if i>1

array_factor = linear_AF(N,d,w,sig(1:size(sig,1),2),E_pattern);
        if (abs(w_err(i) - w_err(i-1)) < eps) | (array_factor <=
AF_thresh)

linear_plot_pattern(sig,w,N,d,E_pattern,AF_thresh,'half',4,'-');
            break;
        end;
    end;
    if rem(i,SKIP_STEP) == 0                % Plot every SKIP_STEP
iterations

linear_plot_pattern(sig,w,N,d,E_pattern,AF_thresh,'half',4,'-');
        if i == SKIP_STEP
            FIG_HANDLE = eval(FIG);
        if (isunix)
            pause;
        end;
        end;
    end;
end;
% Final weights and betas
W    = abs(w);
beta = angle(w);
iterationnumber=i;
%------------------------------------------------------------------
function   [N,d,sig,noise,type,nn,NN,AF_thresh,Mu,E_pattern]   =
linear_data_entry

%%%%%%%%%%%%%%%%%%%% Default Values %%%%%%%%%%%%%%%%%%%%%
def_N              = 8;
def_d              = 0.5;
def_SOI            = 1;
def_q              = 1;
def_SNOI           = 1;
def_noise_mean     = 0;
def_noise_var      = 0.1;
def_type_n         = 2;
def_AF_thresh      = -60;
```

```
def_Mu              = 0.001;
def_nn              = 500;
def_E_pattern_file = 'linear_isotropic.e';
%%%%%%%%%%%%% Strings initialization %%%%%%%%%%%%%%%%%
N_string            = sprintf('Enter number of elements in linear
smart antenna [%d]: ',def_N);
d_string            = sprintf('Enter the spacing d (in lambda)
between adjacent elements [%2.1f]: ',def_d);
SOI_string          = sprintf('Enter the Pilot signal (SOI) ampli-
tude [%d]: ',def_SOI);
SOId_string         = sprintf('Enter the Pilot signal (SOI) direc-
tion (degrees between -90 and 90): ');
q_string            = sprintf('Enter number of interfering signals
(SNOI) [%d]: ',def_q);
noise_mean_string   = sprintf('Enter the mean of the noise [%d]:
',def_noise_mean);
noise_var_string    = sprintf('Enter the variance of noise [%2.1f]:
',def_noise_var);
type                = sprintf('Type of signal:\n\t[1] sinusoid\n\t[2]
BPSK\nEnter number [%d]: ',def_type_n);
nn_string           = sprintf('Enter the number of data samples [%d]:
',def_nn);
AF_thresh_string    = sprintf('Enter AF threshold (dB) [%d]:
',def_AF_thresh);
Mu_string           = sprintf('Enter a value for Mu of LMS algorithm
(0 < Mu < 1) [%5.4f]: ',def_Mu);
E_pattern_string    = sprintf('Enter element pattern filename (*.e)
[%s]: ',def_E_pattern_file);
%----------------------------------------------------%
   %%%%%%%%%%%%%%%%% Error Messages %%%%%%%%%%%%%%%%%%%%%%
err_1               = sprintf('\nSignal type not supported...');
%----------------------------------------------------%
%%%%%%%%%%%%%%%%%%%% User Inputs %%%%%%%%%%%%%%%%%%%%%%%
% ----------- Number of elements? ------------- %
N = input(N_string);
if isempty(N)
   N = def_N;
end;
% ----------- Inter-element spacing? ---------- %
d = input(d_string);
if isempty(d)
   d = def_d;
end;
% -------------- SOI amplitude? --------------- %
SOI = input(SOI_string);
```

```
if isempty(SOI)
   SOI = def_SOI;
end;
% -------------- SOI direction? ---------------- %
SOId = [];
while isempty(SOId)
    SOId = input(SOId_string);
end;
% ------ SNOIs amplitudes and directions? ------ %
q = input(q_string);
if isempty(q)
   q = def_q;
end;
sig = [SOI SOId];
for k = 1 : q,
    SNOI_k_string       = sprintf('Enter the amplitude of No. %d
Interference signal (SNOI_%d) [%d]: ',k,k,def_SNOI);
    SNOId_k_string      = sprintf('Enter the direction of No. %d
Interference signal (SNOI_%d) (degrees between 0 and 90): ',k,k);
   SNOI_k = input(SNOI_k_string);
   if isempty(SNOI_k)
      SNOI_k = def_SNOI;
   end;
   SNOId_k = [];
   while isempty(SNOId_k)
       SNOId_k = input(SNOId_k_string);
   end;
   sig = [sig; SNOI_k SNOId_k];
end;
% ----------------- Noise data? ---------------- %
noise_string = input('Insert noise? ([y]/n):','s');
if (isempty(noise_string) | noise_string == 'y')
   noise_mean = input(noise_mean_string);
   if isempty(noise_mean)
      noise_mean = def_noise_mean;
   end;
   noise_var = input(noise_var_string);
   if isempty(noise_var)
      noise_var = def_noise_var;
   end;
   noise = [noise_mean noise_var];
else
    fprintf('-----------No noise is inserted.-----------\n');
   noise = [];
end;
```

```
% ---------------- Signal type? --------------- %
type_n = [];
if isempty(type_n)
    type_n = def_type_n;
end;
switch type_n
case 1
    type = 'sinusoid';
    def_NN = 100;
case 2
    type = 'bpsk';
    def_NN = 1;
otherwise
    error(err_1);
end;
% ------------ Number of data samples? --------- %
nn = input(nn_string);
if isempty(nn)
    nn = def_nn;
end;
% ------- Number of samples per symbol? -------- %
NN = [];
if isempty(NN)
    NN = def_NN;
end;
% --------------- Mu for LMS? ---------------- %
Mu = input(Mu_string);
if isempty(Mu)
    Mu = def_Mu;
end;
% --------------- Nulls depth? --------------- %
if q==0
    AF_thresh = def_AF_thresh;
else
    AF_thresh = input(AF_thresh_string);
    if isempty(AF_thresh)
        AF_thresh = def_AF_thresh;
    end;
end;
% -------------- Element pattern? ------------- %
E_pattern_file = [];
if isempty(E_pattern_file)
    E_pattern_file = def_E_pattern_file;
end;
```

```
E_pattern = load(E_pattern_file)';
warning off;
f     u     n     c     t     i     o     n
linear_plot_pattern(sig,w,N,d,E_pattern,AF_
limit,TYPE,rticks,line_style)
k0 = 2*pi;
%%%%%%%%%%%%%%%% Parameters initialization %%%%%%%%%%%%%%%%
%---- default values ----%
def_d = .5;
def_E_pattern = 1;
def_AF_limit = -40;
def_TYPE = 'half';
def_rticks = 4;
def_line_style = '-';
H_FIG = 1;
%-----------------------%
switch nargin
case 2

d = def_d; E_pattern = def_E_pattern; AF_limit = def_AF_limit;

T                     Y                     P                     E
= def_TYPE; rticks = def_rticks; line_style = def_line_style;
case 3
    E_pattern = def_E_pattern; AF_limit = def_AF_limit;

T                     Y                     P                     E
= def_TYPE; rticks = def_rticks; line_style = def_line_style;
case 4
    AF_limit = def_AF_limit;

T                     Y                     P                     E
= def_TYPE; rticks = def_rticks; line_style = def_line_style;
case 5

T                     Y                     P                     E
= def_TYPE; rticks = def_rticks; line_style = def_line_style;
case 6
    rticks = def_rticks; line_style = def_line_style;
case 7
    line_style = def_line_style;
end

m = linspace(0,N-1,N);
switch TYPE
```

```
case 'half'
    theta_1 = -pi/2; theta_2 = pi/2;
    theta = linspace(-pi/2,pi/2,181);
case 'full'
    theta_1 = -pi; theta_2 = pi;
    theta = linspace(-pi,pi,361);
otherwise
    disp('ERROR => Only two options are allowed for TYPE: half or
full');
    return
end
%%%%%%%%%%%%% Total Pattern [Element * AF] %%%%%%%%%%%%%
AF = E_pattern .* sum(diag(w)*exp(i*(k0*d*m'*sin(theta))));
AF = 20*log10(abs(AF)./max(abs(AF)));
%%%%%%%%%%%%%%%%%%%%%%%%%% Polar Plot %%%%%%%%%%%%%%%%%%%%%%%
figure(H_FIG);
switch TYPE
case 'half'

linear_semipolar_dB(theta*180/pi,AF,AF_limit,0,rticks,line_
style);
case 'full'

linear_polar_dB(theta*180/pi,AF,AF_limit,0,rticks,line_style);
end

%-----------------------------------------------------------%
%%%%%%%%%%%%%%%%%%%%%%%%% Thermal Noise %%%%%%%%%%%%%%%%%%%%%%%
if ~isempty(noise)
    for k = 1 : N
        STATE3 = sum(rand(1)*100*clock);
        randn('state',STATE3);
        noise_data_real = noise(1,1) + sqrt(noise(1,2)/2)*randn(1
,size(s,2));
        STATE4 = sum(rand(1)*100*clock);
        randn('state',STATE4);
        noise_data_imag = noise(1,1) + sqrt(noise(1,2)/2)*randn(1
,size(s,2));
        noise_data = complex(noise_data_real,noise_data_imag);
        s(k,:) = s(k,:) + noise_data;
    end;
end;
%-----------------------------------------------------------%
function [w, error] = LMS(w,Mu,x,d)
error = d - w' * x;
w = w + 2 * Mu * x * conj(error);
```

# Appendix E: Visualyse Professional

## Appendix E.1: Introduction

Visualyse products are well known by its high computation performance for technical excellence in their support for radio spectrum management, in particular interference analysis. Visualyse Professional has used an underlying model that is based on real-world objects; this means that the structure of a simulation is familiar to an engineer the first time he looks at the software. Building new complex analyses is also made easier by use of mobile system Windows interface components which are also familiar to many people.

Ease of use is not the start and end of usability, nor will a simple software package necessarily enhance the productivity. Usable, productive software must do what you want it to do – it must be effective and provide the functionality that is in need.

This last requirement – utility – is the central goal of Visualyse Professional. The object-based design means that it can be adapted to all types of system; the ongoing software development and cross-fertilization from consultancy work mean the new features and enhancements to existing features. Ease of use remains high on agenda –aim is to make using Visualyse a rewarding experience which is both satisfying and enjoyable. Visualyse finds a balance between ease of use and utility that really can boost the productivity (Fig. E.1).

Visualyse is, at heart, an engine for calculating carrier levels, interference levels and noise levels in radio links. It produces C/I, C/N, C/N + I, pfd, EPFD and I/N numbers and statistics for almost any spectrum sharing or interference analysis scenario it can think of?

It allows to define geometry, dynamics and RF characteristics in a 3D environment that includes the earth as a central gravitational body and can also include terrain spot heights, geo-climatic factors and local clutter data. However, this does not adequately capture the full capability of the software.

© Springer International Publishing AG 2018
L.F. Abdulrazak, *Coexistence of IMT-Advanced Systems for Spectrum Sharing with FSS Receivers in C-Band and Extended C-Band*,
https://doi.org/10.1007/978-3-319-70588-0

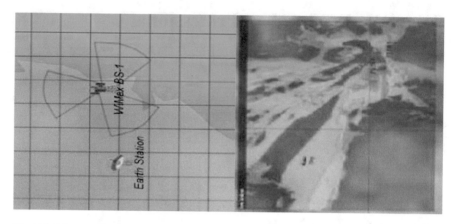

**Fig. E.1** Maps availability in Visualyse software

# Appendix F: Experiment Setup and Minimum Separation Distance Simulation Results

## Appendix F.1: FSS Aspects and Installation

### *Appendix F.1.1: 1.8 m C-Band Antenna System*

**PRODUCT SPECIFICATIONS**

Detail Photos
*(on right from top to bottom)*

Heavy-duty galvanized
Az/El Mount

Fine azimuth and elevation
adjustments

RF tested C-Band Linear
Polarized feed assembly

## 1.8 m C-Band Linear RxTx Class III Antenna System
### TYPE 183

The reflector is thermoset-molded for strength and surface accuracy.

The Andrew Corporation Type 183 1.8 m Class III RxTx Antenna is a rugged commercial grade product suitable for the most demanding applications. The reflector is thermoset-molded for strength and surface accuracy. Molded into the rear of each reflector is a network of support ribs which not only strengthens the antenna, but also helps to sustain the critical parabolic shape necessary for transmit performance.

The Az/El mount is constructed from heavy-gauge steel to provide a rigid support to the reflector and feed support arm. Heavy-duty lockdown bolts secure the mount to any 4.50" (114 mm) O.D.

mast and prevent slippage in high winds. Hot-dip galvanizing is standard for maximum environmental protection.

- One-piece precision offset thermoset-molded reflector.
- Fine Azimuth and elevation adjustments.
- Galvanized feed support arm and alignment struts.
- Galvanized and stainless hardware for maximum corrosion resistance.
- RF tested feed assembly
- Heavy-duty Class III mount for 25 lb (11 kg) RF electronics (LNB & BUC).

© Springer International Publishing AG 2018
L.F. Abdulrazak, *Coexistence of IMT-Advanced Systems for Spectrum Sharing with FSS Receivers in C-Band and Extended C-Band*,
https://doi.org/10.1007/978-3-319-70588-0

**TYPE 183** 1.8 m C-Band Linear RxTx Class III Antenna System

### RF Performance

|  | | C-Band Linear |
|---|---|---|
| Effective Aperture | | 1.8 m (71 in) |
| Operating Frequency | Tx | 5.850 - 6.725 GHz |
| | Rx | 3.400 - 4.200 GHz |
| Polarization | | Linear, Orthogonal |
| Gain (±3 dBi) | Tx | 39.3 dBi @ 6.138 GHz |
| | Rx | 35.4 dBi @ 3.913 GHz |
| 3 dB Beamwidth | Tx | 2.0° @ 6.1 GHz |
| | Rx | 3.0° @ 3.9 GHz |
| Sidelobe Envelope (Tx, Co-Pol dBi) | | |
| 2° < θ <20° | | 29-25 Log θ |
| 20° < θ < 26.3° | | -3.5 |
| 26.3° < θ < 48° | | 32-25 Log θ |
| 48° < θ < 180° | | -10 |
| Antenna Cross-Polarization | | >30 dB (on axis) |
| Antenna Noise Temperature | 10° El | 41°K |
| | 20° El | 36°K |
| | 30° El | 33°K |
| VSWR | Tx | 1.3:1 |
| | Rx | 1.4:1 |
| Isolation | Tx | 60 dB |
| | Rx | 60 dB |
| Feed Interface | Tx | CPR-137 or Type N |
| | Rx | CPR-229 |

### Mechanical Performance

| | |
|---|---|
| Reflector Material | Glass Fiber Reinforced Polyester |
| Antenna Optics | One-Piece Offset Feed Prime Focus Long Focal Length |
| Mount Type | Elevation over Azimuth |
| Elevation Adjustment Range | 7°-90° Continuous Fine Adjustment |
| Azimuth Adjustment Range | 360° Continuous |
| Feed Support | Rectangular Section with Alignment Legs |
| Mast Pipe Interface | 4.50 in (114 mm) Diameter |
| Wind Loading Operational | 50 mi/h (80 km/h) |
| Survival | 125 mi/h (200 km/h) |
| Temperature | -50°C to 80°C |
| Humidity | 0 to 100% (Condensing) |
| Atmosphere | Salt, Pollutants and Contaminants as Encountered in Coastal and Industrial Areas |
| Solar Radiation | 360 BTU/h/ft² |
| Shock and Vibration | As Encountered During Shipping and Handling |

## Appendix F.1.2: System Indoor DW2000 Terminal (DVB Receiver)

Technical Specifications

**Remote Equipment**
Supported Frequencies:
- Ku-band (14.0-14.5 GHz uplink, 10.95-12.75 GHz downlink )
- Extended C-band (5.85-6.425 GHz uplink, 3.625-4.2 GHz downlink)

**Outdoor Unit:**
Ku-band: 0.5, 1.0, or 2.0 watts
C-band: 2 watts

**Data Protocol Support:**
TCP/IP, UDP/IP, IP multicast

**Optional Video Support:**
MPEG 2 and MPEG 1 video decoding

**Optional Hard Drive Size:**
40 Gbytes (minimum)

**Reception Specifications:**
| | |
|---|---|
| Data Rates: | 1.1 Mbps to 24 Mbps |
| Modulation: | QPSK |
| Encoding: | Concatenated Reed-Solomon and Viterbi FEC |
| Code Rates: | Reed-Solomon 130/147; Viterbi 2/3 or 1/2 |

**Transmission Specifications:**
| | |
|---|---|
| Data Rates: | 128 or 256 Kbps |
| Modulation Scheme: | OQPSK |

**Physical Interfaces:**
- Two Ethernet LAN RJ45 ports (10BaseT/100BaseT)
- S-video output for NTSC video
- Three RCA jacks to support composite video (NTSC) and audio (left & right)
- Channel 3/4 RF modulated output to support NTSC
- Two additional RCA jacks for auxiliary audio channels (specific models)
- TVRO output
- Two serial ports (RS-232) for asynchronous traffic (Async to IP conversion)

**Power Supply:**
Universal power supply: 90-240 VAC; 47-63 Hz

**Operating Temperatures:**
Outdoor Equipment:    -30°C to +55°C
Indoor Equipment:    +10°C to +40°C

SNMP Network Manageable using the DIRECWAY Vision Platform

**DiRECWAY.**
HUGHES

### Appendix F.1.3: FSS Unit Installation in Wireless Communication Centre at Universiti Teknologi Malaysia

VSAT Installation started by building a ground mount without fencing to fix the 1.8M C-band antenna on a latitude of 1.33° north and longitude 103.38° east. Then the antenna elevation angle will be 74°, as depicted in Fig. F.1.1.

The DW 2000 Indoor Unit was connected to LNB in order to receive the DVB signal and to a computer to check the internet speed through the RS232 cable. Figure F.1.2 shows the DW 2000 Indoor Unit back connection and the overall indoor setup.

The outdoor unit (ODU) of VSAT consists of the antenna (typically from 0.6 m to 3 m in diameter), equipped with a horn, LNB (low-noise blocking), feed and BUC (block upconverter). The LNB is connected to the receiving loop which con-

**Fig. F.1.1** Antenna dish fixing to MEASAT-3

**Fig. F.1.2** The DW 2000 Indoor Unit overall indoor setup and back connection

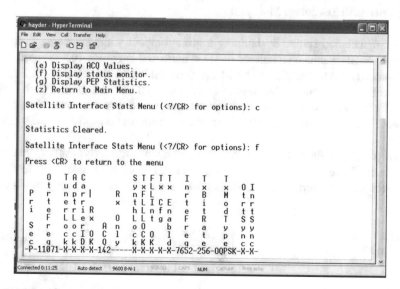

**Fig. F.1.3** The internet speed is 142Kb/s, after the cable and system loss

sists of horn antenna, OMT (Orthomode Transducer) and transmit reject filter, while the transmitting loop consists of BUC (contains a local oscillator and HPA high-power amplifier), OMT and the horn. Note that a circular tube may be added between the OMT and the horn in a case that circular polarization is required. The indoor unit (IDU) is typically composed of a modem to convert the data, video or voice generated by the customer.

The received signal depends on the provider subscription fees; in this research, a bandwidth of 156Kbps downlink and 9.6Kbps uplink was purchased to employ the hardware measurements. Figure F.1.3 depicts the internet speed status.

We conclude that received bandwidth was 230.4 KHz. Figure F.1.4 shows the overall satellite link budget for the FSS unit used for this book.

The indoor unit contains timing units, modulators and demodulators and interfaces to network management systems and host computers for control. To reconfigure the VSAT network, a network management system (NMS) can be used for any dynamic change, like adding more stations and carriers or changing the network interface. Monitoring and controlling the network are part of the operational process inside the NMS. However, NMS should provide a report about each single unit connected to the network. In addition to that, the NMS downloads all the relevant software and system parameters for the system to recall the data in the restart state.

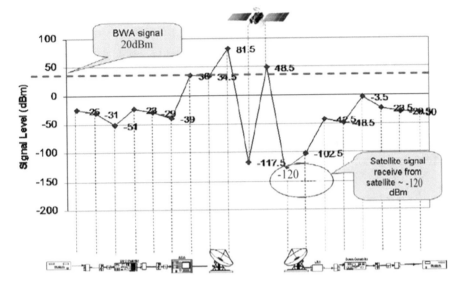

**Fig. F.1.4** Shows the overall satellite link budget for the FSS unit

**Fig. F.2.1** The calibration setup

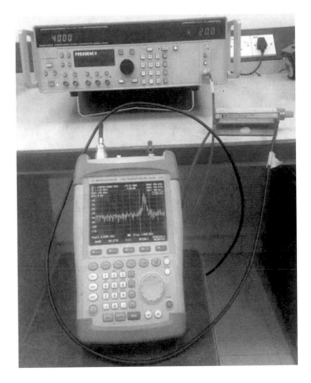

## Appendix F.2: Signal Generator Calibration

Signal generator calibration was done using a handheld portable spectrum analyser and signal attenuator to check the frequency shift and cable loss as well as power level correction. Figure F.2.1 depicts the calibration setup.

## Appendix F.3: Antenna Radiation Pattern and Return Loss Measurement

Radiation patterns are defined as the variation of the field intensity of an antenna as an angular function with respect to the axis. A radiation pattern is usually represented graphically for the far-field conditions in either horizontal or vertical plane. The radiation pattern measurement for the horn antenna which was used for the BWA transmitting signal was performed inside the anechoic chamber located in the

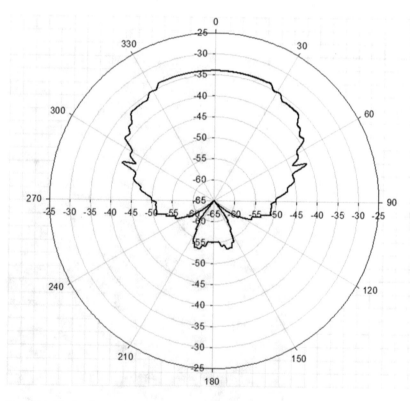

**Fig. F.3.1** H-plan for transmitter antenna

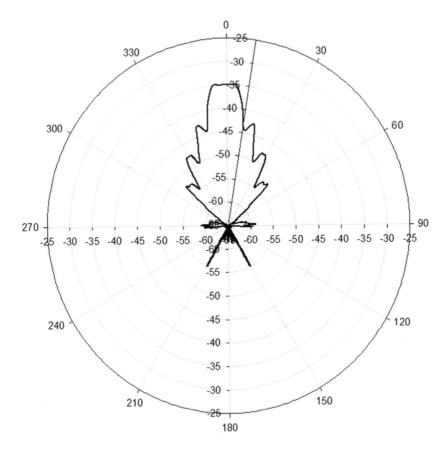

**Fig. F.3.2** V-plan for transmitter antenna

Wireless Communication Centre, Faculty of Electrical Engineering, Universiti Teknologi Malaysia.

An RF anechoic chamber is designed to suppress the electromagnetic wave energy of echoes, as reflected electromagnetic waves, from the internal surfaces. Figure F.3.1 and F.3.2 depict the radiation pattern measurement (H- and V-plan) for the horn. The fixed horn antenna is working as transmitter and in the other side the measured antenna fixed on the rotator base where each 2° rotation the power received measured until it reaches 360° rotation.

The return loss or $S_{11}$ was measured using R&S and FSH handheld spectrum analyser. Figures F.3.3, F.3.4 and F.3.5 depict the $S_{11}$ for the two horn antennas used in experiment test as a receiver and transmitter, respectively

Figure F.3.5 shows the measurement fit-up inside the anechoic chamber with and without shielding to get the radiation attenuation for each material before the outdoor measurements.

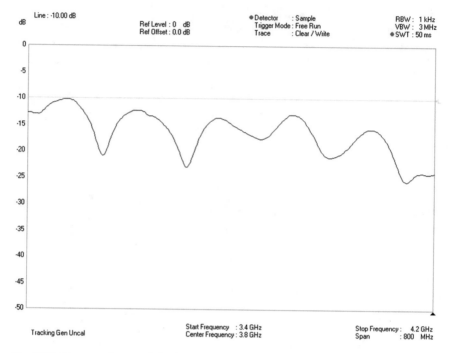

**Fig. F.3.3**  The return loss result for the receive antenna RxS$_{11}$

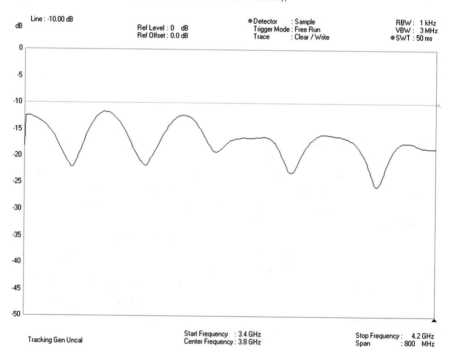

**Fig. F.3.4**  The return loss result for the transmit antenna TxS11

**Fig. F.3.5**  Measurements inside the anechoic chamber with and without shielding

## Appendix F.4: Different Receiving Signals Through Different Shielding

The idea is to measure the amount of signal penetration through different types of materials in order to measure the power loss through several barriers. The shielding is placed on 1 m away from the receiver and 4 m away from the transmitter. Results of deferent receiving signals through different shielding are depicted in Figs. F.4.1, F.4.2, F.4.3 and F.4.4.

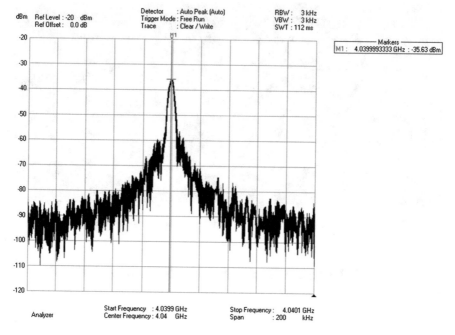

**Fig. F.4.1** Direct signal measurement in the lab

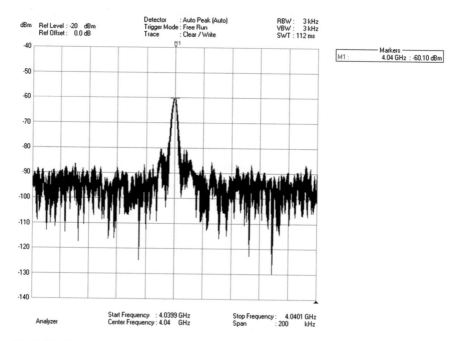

**Fig. F.4.2** Copper attenuation measurements in the lab

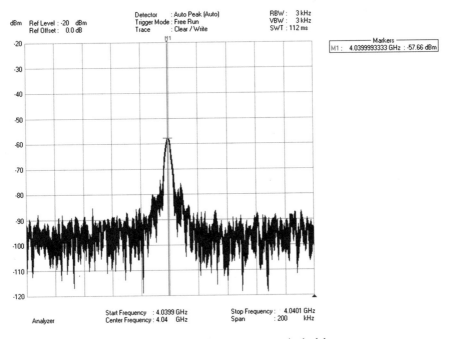

**Fig. F.4.3**  0.1 cm aluminium sheet attenuation measurements in the lab

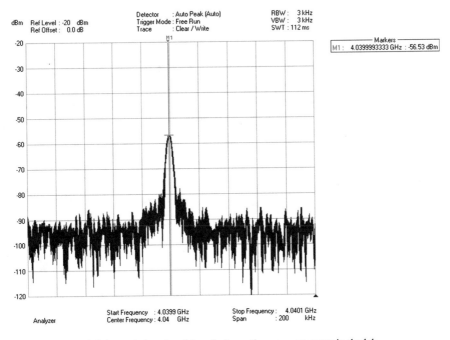

**Fig. F.4.4**  Aluminium mesh (spacing 0.2 cm) attenuation measurements in the lab

# Index

© Springer International Publishing AG 2018
L.F. Abdulrazak, *Coexistence of IMT-Advanced Systems for Spectrum Sharing with FSS Receivers in C-Band and Extended C-Band*,
https://doi.org/10.1007/978-3-319-70588-0